Amphibian men

Amphibian men

Inversion theory of Anthropogenesis

Mikhail Glazunov

First published 2019 by Mikhail Glazunov

British Library Cataloguing in Publication Data

A catalogue record for this book is available from the British Library

ISBN: 978-1-9162532-1-6 (pbk)

ISBN: 978-1-9162532-0-9 (ebk)

About the author

Mikhail Glazunov, Professor of Social, Economic and Political Sciences, of The Russian Academy of Natural History. He is an independent consultant, science writer and a visiting university lecturer. He studied in Russian and British universities and then went on to develop his academic career in the UK and Russia. He is author of the few books on Russian modern history. He is also a contributor to a variety of publications in the modern magazines.

Contents

CHAPTER 7

GENESIS OF CONSCIOUSNESS — 197

CHAPTER 8

APPLIED ASPECTS OF INVERSION THEORY — 225

Preface

The book searches for the predecessors of man based on the unique exclusives of the human body. The scientific method leads to the understanding that our ancestors came not from the wilds of Africa, but from the seashore and were not close relatives of monkeys, but ancient delphinids. This concept is supported by data from different sciences and is difficult to refute. The book demonstrates the possibility of man descending from animals living on the coast and explains the uniqueness of long hair on the head, the absence of fur on other parts of the body, human sweating, how apes lost their tails, the origins of the chimpanzee and other facts.

The book reveals that African skeleton finds are not human predecessors but pathologically altered skeletons of pre-sapiens who left the idyllic coast of Tethys into equatorial Africa ahead of time, the remains of which researchers often optimistically declare as the ancestors of man.

It develops a new perspective on the cause of the origin of consciousness as communication between the different structures of one brain; one can talk about the endogenous origin of consciousness. The fundamental idea of the study of human evolution is the consideration of the phenomenon of human consciousness, not as a result of the slow, progressive development of the animal psyche but as an inversion, a decisive turning-point of the psyche from the animal from which man originated.

When the first steps towards this comprehension were taken, most scientists still believed in the traditional simian theory or in the idea of man as a special project of nature. In recent decades, scientific research has uncovered fascinating findings that upset the simian theory. The book, remaining within the framework of proven facts and rigorous philosophical reasoning, seeks to overturn the reader's view of the origin of man.

Introduction

In 1928 Max Scheler, one of the founders of philosophical anthropology, for the first time systematically posed the question of the special position of man. He separated the two concepts of man: man as a biological entity with a fixed place in the animal kingdom and man as an apparently abnormal being with a specific essence (Scheler, 1928). His approach was discordant with the theory of human evolution, dominant at that time, which postulated that man originated from four-armed monkeys according to the classification of K. Linnaeus. There are various modern variations of this theory, such as man did not come from monkeys, but from a common ancestor, so to designate this concept as a research paradigm it is referred to as the simian theory of anthropogenesis.

Scheler solved the problem of human nature identification theologically, by contrasting the essence of man as the image and likeness of God, with his exceptional animal nature, which only had a chance to be realised thanks to divine support (Scheler, 1966). However, this opposition of *Homo Naturalis* to historical man involved in the divine principle contradicts traditional evolutionary schooling. On the other hand, the idea that the social should be considered not in contrast to the biological, but rather as its product, was actualised in the works of ethologists Lorenz and Tinbergen. They managed to trace the commonality of behaviour between animals and man, but these studies of behaviour did not lead to a solution of the problem of human nature and its origin.

Multiple attempts to identify man and his society with other species such as the lion pride, or a pack of wolves, or an ant colony, or to chimpanzee communities and so on, contains a logical error in substituting the base of the comparison, where the human species is compared with so many all at once. Consequently, there are similarities everywhere, but no complete analogy with any other species. In this case, the comparison is not of holistic human behaviour, but of insignificant parts of such behaviour with that of other species.

Such attempts to equate people and animals produce an inadequate result. However, they lead us to the idea of the special nature of man, whose manifestations of social qualities are truly infinite, in contrast to other species where everything is regulated strictly by instinct. Lions will never be social like chimpanzees, and bee society will not be like that of lions: the drone cannot become the king of the hive and the lioness will not be the queen of her pride.

People can have different societies like the lion pride of the Muslim harem, the polyamorous community of the monkey, monogamous marriages of the swan, or the communistic ant colony. A human leader can be a man or a woman. Or a person can be isolated, entering only into sporadic contact with the aim of satisfying his animal needs and leading him to the limitlessness of asociality, outside the cluster of society. Again, there are analogues in the sporadic contact of tigers, bears and crocodiles.

Sporadic contact of animals for different species is variable. In some species, the initiative is taken by the male and in others by the female, but people of both genders take the initiative to various extents as dictated by culture and personality. Moreover, a man can be a principled hermit who avoids even sporadic sexual contact; even those associated with reproduction, which in animals is impossible in principle. This leads man to a position where the animal property is diminished, except at the anatomical level and behavioural comparisons become secondary to sociological ideas.

Therefore, it can be said that the successes of ethology did not lead to the solution of the problem of bridging the gap between animals and man. They could not answer the question posed by Scheler of how a creature so poorly physically adapted to the wild environment, could survive almost naked whilst leaving a comparatively safe tier of the tropical forest and walking on the ground on short, crooked and weak legs. The last point is especially apt, because initially they were not even legs, but hands; the simian concept postulates the origin of man as being four-armed. Anthropologists, unsuccessfully trying to solve the problem of bipedalism, made their task much easier. They tried to explain why the four-handed human ancestor went on two legs, forgetting that there were no legs as such.

The process of anthropogenesis lasted millions of years. Throughout this time, the belt of tropical forest near the equator was almost invariably preserved. There are tropical species that are more than 60 million years old, such as lemurs and tarsiers. For a long time nothing changed in their existence, they did not need to get down from the trees. It is unclear why monkeys, such as chimpanzees, whose evolutionary age is less than that of lemurs and tarsiers by an entire order of magnitude, could do this. Climate change was not significant in the short period of their evolution. One can see that the problem of the triggering mechanism of the transition to walking upright is not even close to being solved.

Of all the problems of the theory of anthropogenesis, the origins of language and consciousness are the most intractable. Moreover, one can state that there is an absence of an adequate logical approach to them based on the simian concept of human evolution. The fact is that it would be logical to think of psychogenesis and language origin as something procedurally unified with the process of morphogenesis. However, the problems of morphogenesis and psychogenesis are detached from each other conceptually and, even more so, institutionally. Morphogenesis is mainly concerned with biology and palaeoanthropology, whereas psychogenesis is analysed by psychologists and linguists.

Biologists and paleoanthropologists seek solutions to the eternal questions of anthropogenesis such as the trigger mechanism for the transition to bipedalism, the survival mechanism of unadapted human ancestors in the jungle or savannah, the construction of the stages of human evolution starting from hypothetical ape-like ancestors to Homo sapiens. The 'missing link' problem that developed after the discoveries in the middle of the last century, thanks to numerous expeditions has turned into its dialectical opposite: there were too many links between humans and their apelike ancestors.

Accordingly, many lines of anthropogenesis were constructed, but none is conclusive from the point of view of aromorphosis, a progressive evolutionary change in the structure leading to a general increase in the level of organization of organisms. Many features more likely indicate catagenesis, when the older species of evolution lines look more sapient than their supposed descendants. Morphogenesis, as a complex problem, now looks immeasurably more ramified than half a century ago and is much less clear in terms of its resolution. Of the enthusiasm of the 60s and 70s, when it seemed that a missing link was a kind of a keystone which would give an opportunity to complete the whole structure of the theory of anthropogenesis was about to be found, there remains no trace.

Despite the fact that modern anthropologists are evolutionists, they often do not take into account that anthropogenesis is a unified process that includes morphogenesis, psychogenesis and glottogenesis (the development of speech). As a rule, they concentrate their efforts on the formation of human corporeality, leaving the formation of a non-corporeal substance (the realm of spirit) to everyone from psychologists to esotericists. On the other hand, most scientists who deal with the problems of psycho- and glotto-genesis rarely admit, in their research abstracted from human corporeality, judgments about human anatomy and physiology.

If the truly enormous construction of human evolution is not built into something intelligible, the vice has a methodological character. Therefore, it is useful to take a step back in order to analyse the methodology. The error was at the very beginning, when the monkey was arbitrarily assigned as the ancestor of man on the basis of external similarity. The concept of the ancestor should be the result of a scientific investigation, not the beginning. Scheler tried to find a solution, but gave a theological solution rather than a scientifically monistic solution. To search for a methodology capable of proposing a lead to the ancestor of man let us turn to influential researchers that have already become long time classics.

In the 1830s, the founder of modern geology Charles Lyell formulated the Uniformity Principle that declares that past geological events must be explained by causes identical to those now in operation, so the present is the key to the past. The methodology of Lyell gives an idea for understanding the origin of man. If you want to understand how a process was going on, but do not know where it started and how it developed then you can take the end of the thread, without fantasizing about where its beginning lies. The end leads to the beginning, and not vice versa. It is simple, like unwinding Miss Marple's ball of wool. It is not by chance that the perceptiveness of the amateur consulting detective of Agatha Christie's crime novels is associated with the image of the unwinding ball. However, anthropologists of the 19th century did not start from the end, but from what they considered as the beginning. As a result, their efforts have tightened the anthropological jumble into a chaotic tangle, which they themselves can no longer unravel. They started with a monkey, whereas it was necessary to start with a man.

The principle of Lyell means that in evolutionary research it is necessary to isolate the most developed form and go backwards into the depths of time. Other ways of research give more questions than answers; often unscientific and turning into empty fantasies. For example, an amateur palaeontologist found the bones of a fossil creature the size of a dog that had huge, curved claws like a five million years old three-toed sloth. He fantasized for a long time which of the modern animals is a descendant of this being. It is unlikely that he will decide that the ancestor of the horse is before him, but this was the ancestor of a modern horse. If the researcher starts his journey from a modern horse then he can restore the process, which palaeontologists have done, for example, George Simpson in 1951. Modern horses under their hooves have a three-fingered appendage. Then they found a transitional form, which was smaller, where the appendage is longer with no hooves. Then they found one even smaller, with claws - and so on further back in time. Science is unwinding the coil, so often the incorrect scientist tries to push the hypothetical principle he invented himself. The tangle of threads is much easier to unwind from the end.

Philosopher and economist Karl Marx called this method 'the ascent from the abstract to the concrete.' He, like most educated people of the 19th century, thought that man descended from the monkey, and he said that human anatomy is the key to the anatomy of a monkey. It can be noted that this is the path from the end to the beginning, and not vice versa. His theory of socio-economic formations, he based on political economy that is a doctrine of how commodity circulation in the course of its development changes social relations. At the same time, Marx began with the analysis of the formation, where commodity circulation was highly developed, from capitalism, distinguished in it the main specific features, which became the initial principles of the theory. Having created the theory of the development of socio-economic formation, he restored the whole chain of formation up to primitive communism. After creation, the theory was backfilled with content, developed to a level of detail. It all began with a selection of abstractions based on the most developed, actual form. Thus, one can see the progression of the analysis from the final point to the initial.

One can see a similar method in the book of Charles Darwin 'The Origin of Species', where he initially examined the artificial selection applied to man, as the latest and most developed type of selection, then he pulled out the principles of selection and, on this basis, he reconstructed the process of natural selection that nature conducts. Thus, he created a general theory of the origin of species.

There is an obvious difference between religion and science in the methods of cognition. Religion always begins from the beginning, which is a pure declaration, because the beginning is hidden. 'In the beginning was the word', says the Bible (John 1:1), and then follows a chain of unverifiable statements. One can believe, but one cannot verify them. Science begins with the end, namely with the obvious, with the empirical present and goes deep into the theoretical and the past, unwinding the tangle. According to this criterion, the simian theory is rather a scientific religion as it began not from the end (from modern man), but from the unsubstantiated statement that it considered the beginning, and got confused in its contradictions, as many religions are confused. Its apologists defend the four-armed ancestor, without serious arguments, as theologians defend religious dogma.

Nearly one hundred years have passed since the publication of Scheler's 'The Human Place in the Cosmos', but his critical questions are still relevant. How did man evolve from four-handed ancestors? What is his nature?

The wars of the 20th century and experiments focussed upon the study of the limits of the human organism also raised questions about human nature. Absolute dating methods and a technique of the molecular clock developed by physicists and biologists only exacerbated the problem and destroyed relatively well-formed theories so often illustrated by a series of creatures walking in line, when the monkey becomes man, straightening up on the move.

Space research brings forth new approaches. If interplanetary flights are possible, are people on Earth not aliens? There are esoteric hypotheses that treat man as a degenerate descendant of more developed races of intelligent beings who inhabited the Earth in the distant past. The classic approach proceeded from the similarity of man with arboreal apes. Then came the theory of the water monkey, then came even more extravagant theories of the origin of man: from the pig due to the great similarity of internal organs, from the bear due to the similarity of bodies without skins, reptiloid theory and so on.

Russian philosopher and archaeologist Viktor Ten offered to withdraw from psychologism in the construction of the theory of anthropogenesis, to take as the point of reference the corporeality of modern man, and analyse it from an evolutionary point of view without prejudging conclusions based on external similarity. The scientific theory must begin with a methodology, rather than with facts directly, since without methodology the facts cannot create a theory, as facts can be eclectically manipulated. Fact is not evidence but confirmation of proof and a logical, consistent system of evidence can be built only on the basis of a unified methodology.

The traditional theory of anthropogenesis does not begin with the justification of the methodology of scientific research, but with eclectic descriptions of the bone remains of hominids in Africa. In this case, each author builds his own anthropoid evolutionary trees, the multiplicity of which is vivid evidence of the eclectic approach. In this regard, it is reasonable to use the above principle of actualism justified by Lyell, since the natural history of the Earth is the same evolutionary concept as the theory of anthropoid evolution: past evolutionary events must be explained by causes identical to those now in operation, so the present is the key to the past.

Ten adapted the principle of actualism to the theory of anthropoid evolution by systematically proposing to identify in the human body such anatomical and physiological exclusives (APE) that distinguish it from monkeys and other animals and to denote the biological specificity of man. The aim was to determine the type of animal man belongs to: the arboreal, marine or some other type. For this, it is necessary to reconstruct the landscape in which the human body was formed, to get an idea of the conditional factors of anthropogenesis. Various authors since the beginning of the 20th century, for example Hardy (1960), have drawn attention to the fact that man is a naked animal and has subcutaneous fat like water mammals, so they suggested that man be considered a descendant of the water monkey. However, the question arises that if we are talking about adaptation to an aquatic environment, then why is the monkey immediately attractive as a predecessor, and why can these exclusive adaptations not be considered objectively?

Ten (2005), on the basis of the proposed methodology, believes that the possible landscape of anthropogenesis is the coastal border of water and land and not the tropical forest and savannah as was thought before. Moreover, the specific nature of man is the absence of specificity; man is a universal animal.

Many human anatomical and physiological exclusives such as fat distribution, nasopharyngeal, nose and ear shapes, bending of the foot and other signs, sharply distinguish man from hypothetical aquatic monkeys, who would have a completely different type of adaptation to the aquatic environment, characteristic of terrestrial animals. Therefore, Ten hypothesizes that man's distant ancestors were marine mammals. That is, man are animals of marine etiology, adapted to the terrestrial way of life as opposed to aquatic apes, which are terrestrial etiology animals, adapted to a semi-aquatic life.

It is also important to determine the causal factors of anthropogenesis, where the most important is the place in the food chain. The hunting, labour and necrophagy that are offered in the framework of modern traditional simian anthropology are more than controversial. Human predecessors could not be effective hunters because they could not stand in competition with species more adapted for hunting. Manifested as an instinct, labour as a causal factor of evolution leads species in such a narrow deadlock of specialization, where there is no possibility of the occurrence of consciousness. Scavengers have a specific phenotype and endocrinology, which is absent in humans. This book offers an original solution within the framework of the inversion theory of anthropogenesis.

The fundamental novelty of the inversion theory lies in the fact that it links morphogenesis and psychogenesis into a single entity. The rationale for morphogenesis is the foundation of the theory, but the main element is the inversion theory of psychogenesis. This states that consciousness appeared as a result of inversion, an evolutionary dialectical leap (discontinuity of the continuity of quantitative changes), rather than as a result of the step by step complication of monkey reflexes which is impossible, as proved by neuroscience during the 20th century. A possible reason for the inversion is the malfunction of the harmonious animal psyche of pre-sapiens into a schizophrenic state. The anatomical basis for this malfunction was the situation when there were two autonomous brains in one organism, which man inherited from his ancestors.

From the evolution point of view this was a step backwards, an inversion, because adaptability has sharply decreased, pre-sapiens were on the verge of self-destruction; a step comparable to charging the AK-47, when the shutter is retracted before the shot. Due to a split consciousness, thanks to society, consciousness formed. Society here is the main causal factor of the occurrence of consciousness, but this is not the same as the instinctual sociality of so-called social animals. The instinctual sociality of animals is always uniform and unchanging, whereas human sociality is a vector directed towards ongoing improvement. Therefore, the theory presented here applies not only to the past, but perhaps also to the future.

The structure of the book

This book consists of eight chapters. Chapter 1 reviews the main perspectives on human evolution from the traditional anthropological approach. A historical overview shows that palaeoanthropology has unearthed a huge number of artefacts over the last 150 years, and was able to synthesize the approaches of different sciences. The discoveries of new fossils recently and reconstruction of their locomotion and paleoenvironment have significantly increased human knowledge about the morphology and evolutionary relationships of primates. According to the conventional simian approach, the main events of human evolution were the transition of the Hominoid apes from a predominantly arboreal to a terrestrial existence in the open landscapes and the development of bipedal locomotion. It is believed that hominoid transition to life in the open landscapes gave them the need to adapt to an unfamiliar environment, pushed them to seek new ecological niches and stimulated the developmental activity of tools and sociality. The supporters of simian theory consider that human evolution is not a simple and linear process; within the evolutionary paths were bizarre twists, branches and deadlocks. Some of the human populations scattered across the planet became stuck in their development and some degraded. Since the formation of modern humans was a long-term process but evolution is continuous, between the apes that lived 10 million years ago and modern man there were many species and it is possible to talk about the many missing links. However, new kinds of fossil hominids continue to be uncovered with true regularity and there are too many intermediate links and finds are so numerous that their multiplicity becomes a problem for anthropologists. It is clear that with each discovery, with every new described species, the picture becomes more and more complex. The question of the trigger mechanism of anthropogenesis is raised, and may be formulated as why some species of hominoids managed to start using their intellectual and cultural potential much more actively than others, having had approximately the same abilities and anatomical and physiological features of the nervous system.

In Chapter 2, an analysis of the traditional approach is proposed. The main problem of anthropogenesis cannot be considered as solved, even in general terms, because it cannot explain the origins of bipedalism, culture and intellect. Moreover, the morphological approach has revealed the discrepancy between the time of formation of the working hand, the brain and bipedalism and inclusion of the labour factor in the evolution of hominids, taken as the cause of these attributes. According to archaeology, there is no connection between the beginning of stonecraft and the development of intellectual abilities. The traditional approach has not been able to explain convincingly why simian ancestors of humans needed labour as the main driver of evolution. Therefore, one can suggest human labour as a causal factor in the evolution of Homo sapiens that could express themselves only in the form of art, and immediately became creative. This means, as a minimum, the connection of two abstract concepts. Hence, at the beginning of human labour were not Olduvai pebbles but some two-part tool. The many paleoanthropological discoveries do not give highly anticipated answers to the most insistent questions such as: which hominid line leads to Homo sapiens? What was the trigger mechanism of anthropogenesis? How and why there was a transition to an upright posture? How did consciousness originate? The theory of anthropogenesis came to the point where the study of particulars does not give new knowledge because there is a large gap between discoveries in palaeoanthropology and their interpretations.

Chapter 3 presents different arguments both for and against the aquatic ape theory (AAT). Despite the fact that there are many arguments supporting the aquatic ape theory, they are not adequately influential to subdue the opposing arguments. Supporters of the Wading Hypothesis believe that wading through shallow water was a major factor of the locomotion of the earliest proto-hominids that took habitats where considerable wading locomotion was favourable; the differences between the postcranial anatomy of the earliest bipedal hominids and modern humans could be explained by particular adaptations for wading. They also consider wading or swimming resulted in significant hair loss because it increased efficiency and speed due to drag reduction as the absence of body hair improves swimming performance. The better adiposity than other primates can be explained by selection for better buoyancy in water.

However, the AAT does not have an adequate amount of testable hypotheses and associated prognoses and is unclear in both absolute and relative dating. It emphasises various subjectively chosen mysterious Homo-sapiens characteristics and offers purely theoretical explanations for them without any attempt to provide comprehensive, causal explanations for the origins of these qualities. It presents a no more complete consideration of the information about human evolution than researchers already have.

Chapter 4 reveals Scheler's concepts of philosophical anthropology, where he systematically posed the question of the special position of man. The theories of Marx and Darwin are employed for the creation of a methodology of the theory of anthropogenesis. The chapter develops the principle of Lyell's actualism where past evolutionary events must be explained by causes identical to those now in operation, so the present situation is the key to the past. It takes the corporeality of modern man as the point of reference for the construction of the theory of anthropogenesis, and analyses it from an evolutionary point of view without prejudging conclusions based on external similarity. A new approach distinguishes, as far as possible, the features of the human body that differentiate man from animals, and thus investigates where and what the origins of human nature were. These simple features, the anatomical and physiological exclusives, should lead researchers to the landscape in which man and common ancestral species were formed, as well as the factors of evolution which worked upon the creatures from which man evolved. This innovative approach contradicts traditional anthropology that seeks characteristics of current humans based on those of the past, with the purely speculative construction that the answer to the future is in the human past, so going from ape to man created a kind of ideal, constructed primordial creature that is half-human and half-ape, and called it the missing link. The chapter analyses the different anatomical and physiological exclusives such as a special type of scalp hair growth, sweat glands of a special type, cold sweat, special characteristics of the human ear, specific melanocyte cells in the skin, the type of accumulation of fat of the marine mammals and human hypersexuality.

Chapter 5 talks about certain settings in search of our ancestors and attempts to define a more or less accurate historical timeline of when, and where, humanity was born. Based on a number of conditions such as that human ancestors had to be universal creatures, able to live almost anywhere, to conduct a water-terrestrial life and that the human body was formed in a moderately warm and wet atmosphere in salty water. The chapter presents a possible scenario of the formation of mankind. It is assumed that in human evolution could have been a period when the development proceeded on the basis of paedomorphosis and so the ancestor of man could be a neotenic form of Acrodelphinidae. Due to irregularity of the landscape, when the Tethys regression began, numerous groups of Acrodelphinidae began to fall into some bodies of water that have no connection with the ocean. Evolution could offer to lock archaeocetes only to paedomorphosis development through the neotenic form. Neoteny in this case meant that their cubs were able to feed onshore and in shallow water. At the beginning of the regression of Tethys, Acrodelphinidae already had some adaptation to the shallows so a reduction of the water area was a positive factor for their evolution. In addition, their morphology became close to their distant ancestors living at the intersection of two elements: the offspring of the ancient whales had round heads on mobile necks and limbs for a long time, as evidenced by the development of embryos of modern dolphins. The shallowing of the seas and the return to land led to the fact that both hemispheres, both practically autonomous brains, were able to sleep and stay awake at the same time. They began to conflict with each other, which created the conditions for the onset of consciousness. Numerous fossil hominids, which were considered the ancestors of modern man, can be regarded as descendants of pre-sapiens who went deep into the continent too early, before they acquired presence of mind. In turn, today's living descendants of some early hominids such as Australopithecus are modern anthropoids (chimpanzees, gorillas, orangutans and gibbons).

Chapter 6 considers the traditional research paradigm of the origin of consciousness and is based on a step-by-step logic of the formation of human society, when this complex form of existence supposedly arises directly from a less complex form through gradual development. In psychology, consciousness is the main subject in that it is independent of how the attitude to the problem of consciousness is declared in a particular school. The development of the science can be represented in the form of successive attempts to find the origins of consciousness and to determine consciousness through its dialectical opposition; therefore, the subject of research in different paradigms can be imagined as a set of dichotomies of consciousness and its dialectical antithesis. A historical review shows that there are several major problems associated with consciousness that are considered as the mental-physical, mind-body and mind-brain dichotomies, to which one can cite most points of view on the nature of consciousness.

Chapter 7 reveals a new approach, which received a rigorous scientific explanation recently and uses the theoretical synthesis of different scholars such as Pavlov, Kretschmer, Lévy-Bruhl, Porshnev, Chertok and Bekhtereva that negate the logic of the gradual formation of human consciousness and the recognition that it arises from a no less complex form by inversion. The chapter investigates the causes of the break in gradualism, when the development of the genus *Homo sapiens* went against the natural determinants, leading to the appearance of the human species due to abnormal rather than natural selection. Since evolution has no formal logic but rather a dialectic logic, where nature is preserved and changed through its dialectical interplay with itself in order to develop, it does not stagnate in the equilibrium of highly specialized species.

A fundamentally new approach is offered to the endogenous theory of the origin of consciousness that suggests that in the course of evolution of the ancestors of man, two relatively independent brains were consolidated into one organ, actively communicating, and competing within it. However, at the point of formation of the human brain, there was still no agreement between the two parts, so there is insanity; our ancestors were schizophrenic and autistic. Many populations of pre-sapiens died, as they were unable to withstand this mental breakdown. Only a small group of people, apparently by the consent of the love of a woman, could survive.

Chapter 3 considers the application of the inversion theory of the origin of consciousness to certain problems of psychology, art and linguistics. The chapter reveals the phenomenon of distinct trance states that are usually accompanied by visionary activity that, as empirical descriptions, are widely present in literature and art. Visions are inextricably linked with the development of mankind, which ascribes specific cultural and historical meanings to the visionary. The inversion theory gives a more coherent explanation for the phenomenon of synaesthesia, which is the stimulation of one sensory modality dependably causing perception in one or more different senses. The hypothesis that primitive people had a cross-sensory perception that existed simultaneously with split thinking is presented. At the initial stage of evolution, the primitive man had pralogic thinking and stable cross-modal transitions of sensory pathways in the brain. In the process of psychogenesis, man has learned to share modalities and independently analyse them. Within the framework of the endogenous theory of the origin of consciousness, synaesthesia was present as a permanent neurophysiological reaction in both hemispheres and ensured the formation of a complex polymodal image of perceived objects in primitive man. That improved survival chances but also brought dangers of its own, like a game of contradictory images in the brain. The chapter investigates the metaphor that is understood not from a narrowly semantic analysis of the artistic pathway but widely from the measurement of cognitive constructs that are part of the integrative processes of subjective experience.

Any inquisitive person would be interested to know the truth about himself, about the secrets of the structure of the human body and psyche.

The concept offered in the book allows us to discover the secret of the beauty of the human body in a new way. In addition, it significantly expands the horizons of thinking, freeing it from many stamps. The book offers a fundamentally new concept of human origin and it completely changes the traditional notions of anthropogenesis. To substantiate the concept, the author draws many bright and interesting scientific facts, considering them from an unexpected viewpoint. He does not bypass the difficult questions of the theory of anthropogenesis, and moreover raises new questions of anthropogenesis and psychogenesis. A new theory of human origin based on a systemic approach and dialectics is so firmly built that it is difficult to imagine the possibility of a competent refutation.

The book is designed to be understood by a broad range of readers: biologists, psychologists, professional philosophers, university lecturers and students who are interested in new approaches in anthropology or who want to escape from old-fashioned anthropological models. In addition, it may interest readers who are attracted to the problem of the descent of man and human development. The book provides a useful tool for understanding inversion theory prospects today, giving the reader new methodological approaches for the development of anthropology and other sciences and so allows the reader to find new discoveries.

Chapter 1

The traditional approach to human evolution

In search of human progenitors

The idea of great similarity of human and monkey was not alien even to scholars of antiquity starting from Aristotle; it is contained in the writings of the Roman physician Galen who widely used baboons in anatomical studies. In 1735, Linnaeus placed man in his classification near apes and gave no religious views of human origins. It began to assert the idea that evolution applied to human beings. Jean-Baptiste Lamarck admitted the possibility of the origin of man from an ancient ape, which has moved to a new way of living due to changes in the environment. Later, Dutch anatomist Camper showed a profound similarity in the anatomical structure of the main organs of humans and animals and a research of French archaeologist Boucher de Perthes proved the existence of primitive man who lived at the same time as the mammoth (Cohen & Hublin, 2017).

A Neanderthal skeleton found in 1856 exploded the scientific world and pushed researchers to the organization of field missions, the creation of a new science - palaeoanthropology and to raise the question of the biological evolution of man in past epochs. An important role in the development of the simian theory of anthropogenesis, that is to say the origin of man from the fossil ape, fell to the ideas of Charles Darwin. In his book, 'The Descent of Man and Selection in Relation to Sex' (1871) he collected a huge amount of evidence of the animal origin of man from comparative anatomy and physiology, embryology, pathology, archaeology and palaeontology. Darwin also noted common emotions of humans and monkeys and ways of their expression, in his 'Expression of the Emotions in Man and Animals' (1872).

Charles Darwin was faced with the fact that by his own theory of natural adaptive selection human evolution cannot be explained. Anatomy and physiology point to the non-adaptive factors shaping many features, and sometimes even anti-adaptive factors. He noted that no one would think that the absence of hair on the skin brought direct benefit to humans; so the hair could not disappear from the body by natural selection and additionally there is no reason to think that it could happen due to the direct impact of climate change. However, he limited his research to a narrow range of issues, such as the lack of hair on the body and its development on the face and head, the mental and physical advantage of men over women, color of the skin and large volume of the brain of humans in comparison with that of animals.

At the core of anthropogenesis, Darwin put sexual selection rather than natural adaptive selection, which explains the origin of all other species. He came to the conclusion that of all reasons that led to the external differences between man and animals, sexual selection was the most active, moreover he presumed that man evolved from some ape-like ancestor and subsequent morphological changes can be explained by sexual selection.

Darwin built his basis of such judgements on the assumption that in the lowest parts of the animal kingdom, sexual selection had no apparent effect. Such animals often stay for a lifetime in one place or in some cases both genders are present in a single individual. Otherwise, more importantly, their senses and mental faculties are not developed for love, jealousy or free choice. However, when we rise to the anthropoids and vertebrates, we see already significant impact of the sexual selection.

Darwin's famous book develops the issue of evolution in the already formed species of *Homo sapiens* and, as a result, the anthropogenetic ideas contained in it have not developed appropriately. Despite the fact that Darwin never said that man descended from apes, in the public consciousness he is the author of 'the theory that man evolved from apes'. Due to the similarity of humans and monkeys, the concept of human origins from ape ancestors in the Darwin's time and after was the most popular. However, Darwin did not prove that sexual selection can form a new species and did not find a solution for the problem of transfer of the method of sexual selection from one class of animal to another.

In support of the anthropogenesis mechanism, the idea proposed by Charles Darwin of increasing the role of sexual selection from lower to higher taxa does not work in general terms. The factor of sexual selection has the highest role in birds who are the direct descendants of dinosaurs. Birds are a parallel line of evolution; the discrepancy of evolutional lines between birds and mammals occurred at the level of reptiles or perhaps at the level of fish. Mammals have much less sexual dimorphism in comparison with birds. It is often difficult to distinguish males from females externally; often, gender identity is only possible on the external genitalia.

The bird kingdom provides many examples of strong influence of sexual selection, but to extend them to man was not possible and Darwin could not find examples of how sexual selection can affect the change in anatomy of mammals in the direction of the creation of Homo. The Darwinian idea that the beard appeared because one of the men was liked by females and so the beard was inherited by his descendants in the male line cannot be accepted as evidence of sexual selection.

Darwin stated that the fossil ancestors of humans lived on the African continent. He believed that mammals living in every major region of the world are closely related to the fossils in the same region. It is possible that in Africa in the past lived now extinct apes, akin to the gorilla and chimpanzee. Because these two types are closest to man, it is somewhat more probable that our early predecessors lived on the African continent than elsewhere.

Evolutionary anthropology was also developed by philosophers such as Engels in 1876 in 'The Part Played by Labour in the Transition from Ape to Man'. He revealed the specifics of the labour process of anthropogenesis, distinguishing the appearance of Homo sapiens from the genesis of other species. Engels suggested a heuristic triad: the brain, bipedalism and working hand, the formation of which formed in the triune process according to principle of positive feedback. He believed that the humanization of apes began with the fact that they moved to an upright posture. Consequently, their hands became more free and they began to work, so step by step hands became more skilful and able to craft elementary tools. Consequently, he highlighted that the hand was not just the organ of labour; furthermore, it was the product of labour. Because of this, the brain began to grow and develop. The development of the mind in turn complicated labour operations that require a more and more straightened position. So formed the human triad where the main condition is trinity and interference factors (Engels, The Part played by Labour in the Transition from Ape to Man, 1934).

The scope of the family *Hominidae* depends on what attributes are used to determine the place of specific species in the superfamily *Hominoidea*. Since man is the single modern representative of the family, his features have historically been divided into three major systems that are considered truly hominid: bipedalism, adaptation to operate with hand tools and a highly developed brain.

Thus, philosophers and anthropologists proposed an ape theory, as they tried to find answers to questions of what prompted men to use the available cultural potency more actively than did the great apes, which possessed roughly the same abilities and have the basic forms of culture, sharply distinguished from monkeys and other mammals. Despite the passage of several million years, the apes did not receive any significant development in comparison with man. For some reason, the use and manufacture of tools, symbolic communication and other elements of cultural behaviour ceased to be for our ancestors something random and sporadic but acquired critical importance for their existence. Why, in other words, has the cultural Rubicon been crossed that separated the people from their animal cousins? This question is the very essence of the problem of human origins.

Some researchers explain this by the rapid development of the hominid's brain, which allegedly allowed them to embark on the path of humanization. However, in fact, this explanation does not address the problem of understanding why, from several species of monkeys having the same size of brain, some of them suddenly became able to think whereas other continued to use the brain poorly. In the first few million years of evolution, the hominid brain was still quite like that of a monkey and exceeded neither the size nor the complexity of the structure of modern gorillas and chimpanzees, but later it began to grow and become more complex.

The development of the idea of evolution and discoveries in geology and archaeology in the nineteenth century encouraged naturalists to search for a 'transitional link' between man and ape; or more precisely the ancient ancestor pongids. Discoveries were often made accidentally but later, expeditions began to be organized and the search for human ancestors became every year more systematic and widespread. Moreover, palaeoanthropology attracted knowledge and scientific technology from various sciences such as geology, biology and physics.

Anthropogenesis (the study of the origins of man) is the branch of anthropology that studies the origin of man and human variability over time. Often palaeoanthropology is used as a synonym but it also has another, narrower meaning. Sometimes it is understood as the branch of science that studies the ancient people of the modern type. In addition, anthropology in general, and in particular anthropogenesis, is confused with archaeology which actually deals with completely different things. Anthropology in our context (physical anthropology) is a biological science, studying the *Hominidae* that includes great apes, Homo sapiens and several extinct species classified as closely related to modern humans, in their biological and evolutionary dimensions. Archaeology, on the other hand, is a social science that studies artefacts, public relations and social structure. Ideally, anthropology and archaeology should complement each other.

However, before we explain the understanding of the theory of anthropogenesis by the modern simianist, we should consider the basic data at the disposal of the modern anthropologist. Below is a list of the major discoveries made over the past 200 years, which determine the basic theoretical concepts of anthropology.

Great expectations of the 19th century

Red Lady of Paviland

In 1823, William Buckland, Professor of Geology at Oxford University, found an almost completely preserved skeleton painted with red ochre in Goat's Hole Cave, one of the limestone caves of Gower Peninsula in the south of Wales. When Buckland discovered the skeleton, he concluded that he had found a female skeleton belonging to a Roman prostitute because it was painted in a red colour and had perforated seashell necklaces and jewellery painted red. Later, 'the lady' was defined as a man of 21 years old who lived about 33 thousand years ago in the end of the Upper Paleolithic and he is the most ancient human relics of anatomically modern *Homo* in the UK (Jacobi, 2008).

Homo neanderthalensis 1856

In 1829 in Belgium and 1848 in Gibraltar were discovered relatively similar skulls, which were later named *Homo neanderthalensis* in honour of the discovery, in a limestone quarry of the Neanderthal valley near Düsseldorf in 1856. The anatomist Hermann Schaaffhausen made a description of the skull and supposed that it is a fossil of an ancient man; later in 1863, William King named the species neanderthalensis (Murray, *et al.* 2015).

However, German scientist Virchow disputed the fact that the discovery was a fossil hominid, believing that it was the skeleton of a diseased Cossack cavalryman with a particular form of bone caused by rickets and Neanderthal is a modern human; he pointed out that some skulls of modern times showed a similar development superciliary arches and a forehead sloping back nearly as much. However, in 1859 after the publication of Darwin's theory of evolution, anthropologists began to consider *H. neanderthalensis* as intermediate link of transformation from ape to man.

Present views on the Neanderthals are different from those of Darwin's time. Neanderthals were ancient people from the Early Paleolithic. Skeletal remains of Neanderthals have been discovered in Europe, Asia and Africa. They lived between 200 and 28 thousand years ago. Study of the Neanderthals' genetic material shows that they apparently are not the direct ancestors of modern humans. They are regarded as the independent species *Homo neanderthalensis*, but some researchers suppose that they are a subspecies of *Homo sapiens* (*Homo sapiens neanderthalensis*). The majority of the remains of Neanderthals approximately 200 individuals were found in Europe, mainly in France and they lived in the period of 70-35 thousand years ago. According to modern representation, they are a predatory version of *H. sapiens*. Neanderthals had a height of about 165 cm, a massive constitution and a large head. The volume of the cranium, between 1400 and 1740 cm^3, even surpassed that of modern humans. They are distinguished by powerful eyebrows, a prominent nose, a very small chin, short hands and a short neck, which was tilted forward. There is speculation that they could be red-haired and white-skinned.

They have many human features in anatomy and behaviour, but are still markedly different from us including the much more massive skeleton and skull. Probably, many of the features of the Neanderthals in Europe were formed under the influence of harsh conditions of the Ice Age about 70-60 thousand years ago. Interestingly, the brain volume of some representatives of Homo neanderthalensis exceeds the values typical for the modern man. The Neanderthals could interbreed with *H. sapiens* and modern non-African populations of *H. sapiens* have approximately 2.5 per cent of Neanderthal genes (Päabo, 2014).

There are many hypotheses about the causes of the extinction of the Neanderthals such as a literal version of the extermination of Neanderthal sapiens (a kind of ancient genocide) or that the European Neanderthals were unable to compete with more advanced Cro-Magnons, that kicked out Neanderthals from the best places and they withered and died out. Golovanova (2010) offers that perhaps volcanic winter and climatic stress was the reason for the extinction of Neanderthals.

Darwin's ideas and human paleontological findings caused an explosion of enthusiasm among biologists including the German zoologist and evolutionist Haeckel, who proposed the new philosophy of anthropomorphosis (Richards, 2008). In 1874, he published the first ever-comprehensive study of human evolution 'Anthropogenie'. Haeckel rejected Darwin's proposition of Africa as the place where the first hominids were spawned and further suggested that the missing link could be found in the vanished continent of Lemuria, which hypothetically could be placed in the Indian Ocean (Haeckel, 2010). The continent, according to the concept, joined Asia and Africa; it was the home of the first humans and allowed them to move outside of Lemuria. Evidence of human evolution could be discovered in the Dutch East Indies due to that humans are strongly related to the Southeast Asia primates. His ideas fascinated many people, amongst them was Dutch doctor Eugène Dubois.

Cro-Magnon 1868

In 1868, in a rocky grotto in Cro-Magnon in France, French palaeontologist Louis Lartet discovered skeletons together with Upper Paleolithic tools. Later, Cro-Magnon became used as a universal name of the first *Homo sapiens*. Cro-Magnons appeared much later than Neanderthals 40-30 Tya (thousand years ago) and they coexisted with them for some time. In outward appearance and physical development, they were practically indistinguishable from modern man. From an evolutionary point of view, the morphological structure and the complexity of the behaviour of these people are not very different from modern *H. sapiens*; nevertheless, anthropologists note a number of differences in some bones of the skeleton and skull from modern Homo.

The term 'Cro-Magnon' can mean in the narrow sense only people found in the Cro-Magnon area who lived around 30 thousand years ago but in a broader sense is the whole population of Europe or the world of the Upper Paleolithic. Cro-Magnon fossils were found on different continents, for example, Africa (Cape Flats), Asia (Nia), Australia (Mungo) and America (Los Angeles) and in almost every one of these locations was found remains of several individuals, often of different ages and gender; in most cases it was a ritual burial.

Cro-Magnons inherited from their ancestors a highly active brain, so in a relatively short period of time, they managed to make an unprecedented step forward. This is evidenced by the development of communication, manufacture of tools, new forms of social organization, art and active adaptation to external conditions.

Some researchers believe that the Cro-Magnon is the ancestor of all modern humans, appearing in East Africa 100-200 Tya and later, around 50 Tya, they migrated from Africa to the Arabian Peninsula and then to Europe and Asia. According to another version of events, Cro-Magnons were distributed for the most part only in the area of Neanderthals whereas the modern representatives of Negroid and Mongoloid races were formed independently.

1891 Pithecanthropus erectus (Java Man)

Dubois was keen on the idea of Haeckel, about the existence of fossil forms, intermediate between man and apes. In 1887, when he was a doctor of medicine, he decided to abandon his career and go to South-East Asia in search of the 'missing link'. He needed to enter military service, with the rank of sergeant of the Royal Dutch colonial army. For one and a half years, he conducted excavations in caves, similar to the studies of human remains in Germany, in his spare time from the responsibilities of a military doctor. He was making long journeys, explored many caves, but could not find residual ape-men, stone tools and signs of settlement (Swisher *et al.* 2000).

He gathered a rich collection of residues of extinct species such as the southern elephant Stegodon, leptobos buffalo, various deer, hippos, tapirs and so on and he published many reports. The colonial administration paid attention to his work and commissioned him to do paleontological research in Sumatra. As a result, he received funds to conduct excavations and could finally leave military service.

In September 1891, in a river valley in Bengawan (Java), he found a fossil tooth similar contemporaneously to human and ape, which was deposited in rock layers of about a million years of age. After two years of hard excavation, a skullcap and a thighbone were revealed. Dubois thought long about the findings: the skull had primitive characteristics but the volume of the brain cavity was significantly greater than that of any ape and the tibia had characteristics of upright walking (Shipman, 2001).

Careful measurements and comparisons with similar bones of men and apes convinced Dubois that that the fossils represented the missing link between apes and humans, he called this finding *Pithecanthropus erectus,* a name proposed by Haeckel, then later renamed it *Java Man.* Today, they are classified as *Homo erectus.* These were the first specimens of early hominid remains found outside of Africa or Europe (Swisher, *et al.* 2000).

Dubois' fossil bones of Pithecanthropus aroused great interest and was the cause of a long and heated debate. He spook before anthropologists in various European cities, demonstrating his findings. Some anthropologists supported his point of view but others angrily dismissed his arguments and stubbornly refused to recognize the validity of the conclusions of the 'missing link'. In December 1895, in the Berlin Society of Anthropology, Ethnology and Prehistory was organised a conference about the Dubois remains. The numbers of primitive features of the skull such as low sloping forehead and massive supraorbital torus of *Pithecanthropus* led to scepticism in the conference of Dubois' findings of an ancestor of man and Society President Rudolf Virchow said that the skull must belong to a giant gibbon and the femur is not connected with the skull.

For two years he fiercely fought for *Pithecanthropus* and finally, exhausted by insults and attacks by colleagues, he put the Pithecanthropus bones in the store of the Leiden Museum. For twenty-five years, they were in a safe and nobody was allowed to see them. At that time, he did not take any part in scientific discussions and it seems that the fate of the discovery became indifferent to him. Only in 1923, when in China the remains of *Sinanthropus pekinensis* was found, Dubois returned from seclusion.

In 1930, Ralph von Koenigswald discovered other, better-preserved, remains of *Pithecanthropus* (Homo erectus soloensis) on the island of Java, after which the doubts about the results of Dubois and connections of Pithecanthropus to the genus Homo disappeared. Resemblance between Java Man and Peking Man led anthropologists to rename both as *Homo erectus*.

Later, findings were made in Africa, the southern regions of Europe and Asia. Different species of Pithecanthropus have been identified: *P. Mauritanicus* of North Africa *P. Palaeojavanicus* of Java and *P. Pekinensis* in China. Apparently, there were at least two major geographical branches of the hominid: Afro-European and Asian. Modern dating techniques indicate their existence between 1.5 Mya and 0.4 Mya.

A number of researchers combined *H. erectus* into one species with the more ancient *H. ergaster*, but the differences between them are quite large. The later hominid *H. erectus* is distinguished mostly by lower skull and parts of the facial morphology. *H. erectus* was unquestionably man, but very different from modern man, so many anthropologists tend to allocate them as a special kind of Pithecanthropus. They were representatives of the early and middle Acheulean culture.

Global discoveries explosion in the 20th century

Homo heidelbergensis 1907

In October 1907, near Heidelberg in western Germany during excavations in Mauer, a worker found a huge jaw like that of a monkey, but with huge humanlike teeth and this served as the holotype for the species. The finding was transferred to Professor Otto Schoetensack, who identified the pattern and gave it the name. Heidelberg man is a comparatively late form of Homo, close to *H. erectus*. It differs from archanthropes by a large brain and from paleoanthropes by the archaic, low skull with very thick walls and massive jaws with large teeth. The culture of these hominids was mostly the Acheulean, sometimes with signs of the transition to the Mousterian. The age of the findings was determined as 400 thousand years. Later, in Israel (1925), Germany (1933) and France (1971) more *H. heidelbergensis* remains were found (Cartmill, *et al.* 2009).

It is important to note that almost all the discoveries from the 19th century until 1925 were described by their own taxonomic name. At that time a small amount of human fossils were investigated: European Neanderthals and Cro-Magnons and Pithecanthropus of Java and Heidelberg. The evolution of human ancestors was represented as phased transition from one stage to another and appeared as the series: Pithecanthropus, Neanderthals, Cro-Magnons and Homo sapiens. Although even then many anthropologists did not take Pithecanthropus and Neanderthal as human predecessors.

Rhodesian Man 1921

In 1921 in Broken Hill (Northern Rhodesia), miners found a cranium, an upper jaw, a tibia and other bones of few individuals. Arthur Smith Woodward made a description of the specimen and named it as *Homo rhodesiensis*. Based on the analysis of related stone tools and animal remains, it is dated to be between 130 and 300 thousand years old.

Rhodesian man is a mixture of archaic and innovative features. The cranial capacity of his skull has been estimated at 1,280 cm^3; the frontal bone is very primitive with massive brow ridges (Rightmire, 1993). The structure of the preserved bones of the postcranial skeleton is practically modern, the height was estimated at approximately 178-184 cm and the proportions of the body were narrow and elongated (Ibid). Currently most anthropologists define *H. rhodesiensis* as *H. heidelbergensis*.

Australopithesus africanus 1924

In the early twentieth century, despite the fact that most of the finds of fossil humans with large brains were made in Eurasia, Darwin's idea that humans had originated in Africa did not have widespread acceptance from researchers. In 1924, a limestone miner from the small town of Taung in the North West Province of South Africa exposed a breccia-filled cave and and found a child's skull.

The remains come into the hands of Raymond Dart who established that the skull belonged to a six-year-old with a full set of milk teeth. The molars that usually appear in man at the age of six years had only begun to erupt. In addition, the teeth resembled human baby teeth in size, unlike the highly developed canines of other primates. The skull was high and round, with a small front part. The new finding was more like a chimpanzee skull, but the top was too convex and its brain dimensions were too large to be from a chimpanzee. Dart also noted that the foramen magnum, which serves to accommodate the spinal cord within the bottom of the skull, indicated that the six-year-old walked upright on two legs. In higher primates, this hole is closer to the occiput, which is connected with their four-legged walking. Dart concluded that before him was the missing link, a transitional stage from ape to man. Soon Dart's sensational article 'Australopithecus Africanus: The Man-Ape of South Africa' appeared in the journal Nature and the scientific community learned about the australopithecus - the long-awaited 'missing link' between apes and the already known by the time *Pithecanthropus* (*Homo erectus*). Dart called this extraordinary little thing Australopithecus africanus, ie. Southern (Australo ...) monkey (pithecus) from Africa, the southern ape of Africa. However, it quickly became known as 'the Taung child' (Garwin & Lincoln, 2003). In the Taung cave together with the young skull of Australopithecus were found the bones of baboons, antelopes, tortoises and other animals. The skulls of baboons were breached as if by a kind of blunt instrument, therefore Dart suggested that these faunae were remains of the ape-men's feast. He argued that the Australopithecus had been powerful hunters and there was 'predatory transition from ape to man'. So an image of Australopithecus as a skilful hunter and good runner on the savanna who killed baboons by blow of a cudgel to the head was created (Dart & Craig, 1959). However, a detailed study using microelement and isotope composition of bones showed that the accumulation of bones found there are really remains of the banquet not of ape-men but some other predators. Australopithecines were not the hunters but rather victims. Among the possible hunters of ape-men are big cats such as sabre-toothed megantereon and leopards (Bunn, 2007).

Proportions of *A. africanus* were intermediate between the proportions of chimpanzees and modern humans and the dimensions of the body were about one and a half metres and the weight was between 20 and 40 kg. His legs had progressive features, but the big toe probably had a high mobility. Pelvic bones were closer to man than to the pelvis of anthropoid apes, which confirms bipedal locomotion of *Australopithecus*. The slope of incisors and canines and their small size of *A. africanus* are different from anthropoid apes and more like those of a man. The structure of the brain of Australopithecus africanus is more similar to the human brain than the brain of other kinds of Paranthropus but the brain volume of *A. africanus* is sufficiently less than that of modern man. Currently, Australopithecus is believed to have lived 3.5 - 2.4 Mya.

1936 Paranthropus robustus (or Australopithecus robustus)

In August 1936, Dart's closest friend Robert Broom discovered fragments from few hominids in limestone caves of Sterkfontein (Northwest of

Johannesburg) which was named *Plesianthropus transvaalensis*, but later classified as an *A. africanus*. In 1938, a schoolchild found a skull imbedded in rock. Broom, who conducted the excavations at Sterkfontein Caves, bartered it for five chocolate bars. The skull and a number of other fossils in this location were described as *Paranthropus robustus*. They had extremely large molar teeth, large sagittal crests and jaws that were adapted to serve in the dry environment (Broom, 1950). *P. robustus* skull was massive, with thick walls and expressed relief. The forehead is low and sloping. The superciliary roll is powerful and the eyebrows are almost horizontal; the facial skeleton is flat, even concave. The foramen magnum was shifted forward, which is evidence of the bipedalism of the species. The endocast has a primitive form, a little different from the forms of anthropoid apes.

The upper limb of *P. robustus* represents a mosaic of pongid and hominid traits whereas the very broad morphology of the pelvic bones was certainly closer to humans than to apes and the total complex of features suggests their bipedal gait. The average brain size of *P. robustus* was similar to the chimpanzee (around 500 cm^3).

In the place where they found the remains of *P. robustus* was found many bone fragments with traces of wear on the end. Empirically, it has been shown that similar traces are formed when using bones for catching termites; thus, it is possible, *P. robustus* could use bone tools. In general, *P. robustus* is dated to have lived between 2.0 and 1.2 Mya. Thus, *P. robustus* is a variant of Australopithecus. Broom's discoveries of A. africanus and *P. robustus* showed that the evolution of Homo was not a straight line and had huge divergence.

In the second quarter of the 20th century, numerous discoveries were made of completely new species of fossil Hominidae, such as Australopithecus africanus and Prometheus, Plesianthropus transvaalensis, Paranthropus Robustus, Paranthropus Crassidens and Pithecanthropus pekinensis (McCown & Keith, 1939; Broom, 1950). Practically every new finding was described as a separate genus and species name. Schemes became more complex and numerous branches were introduced. In describing the findings, special characteristics were often emphasized and the concept of stages of anthropogenesis was settled.

New technology and the growth of contradiction

Dating of the bones of ancient people

In 1949, Willard Libby from the University of Chicago revolutionized archaeology, inventing a method of radiocarbon dating of wood, bone and other organic materials under the age of 40,000 years. Libby's invention was a direct consequence of the development of the atomic bomb in the USA. At first, he tested his method on organic objects, whose age was already known, such as Egyptian mummies, but soon he began to date archaeological sites, which were inhabited for thousands of years before. Today, radiocarbon dating, based on the use of accelerator mass spectrometry (IMR) of radiocarbon, allows determination of the age of such small objects as a single grain on an ancient farm.

Developed in the 1970s, the potassium-argon method is used for chronometric dating of prehistoric archaeological sites with an age of between 50,000 and 2 billion years (Dalrymple & Lanphere, 1970). Since many groups of early people lived in limestone caves and rock shelters, the bones and artefacts became embedded in layers of calcium carbonate and therefore can be dated by a radiometric dating technique. This is a method based on the decay of various uranium isotopes, which decay through a sequence of stages to stable isotopes of lead. This method can be useful for dating items originating between about 10,000 and 2 million years ago.

Australopithecus Boise 1959

In July 1959, in the famous Olduvai Gorge (Tanzania), Mary Leakey found the well-preserved cranium of Australopithecus that had a large and robust constitution and was dated from around 1.8 Mya. The finding got its name in honour of the British executive Charles Boise, who helped to finance paleontological research in East Africa; initially *Australopithecus boisei* was called *Zinjanthropus boisei*, where Zinj is taken from the Old Persian geographical name of the East African region (Leakey M. D., 1984). Some paleoanthropologists believe that *A. Boise* is not an independent species and represents only a local variety of *A. robust*. Nevertheless, it is characterized by many distinguishing features. According to the reconstruction made by studying the fossil bones of the skull and limbs, some specimens of this species achieved the height of modern man. They had a cranial capacity of around 500 cm³ (Dean, 2000), similar to modern apes. They walked upright and their powerful constitution looked like that of a gorilla and had remarkable sexual dimorphism. Like the gorilla, *A. boisei* had a large skull with supraorbital ridges and a central bone ridge, which served for the attachment of powerful jaw muscles. These Australopithecus achieved substantial development of the jaw apparatus and due to the size of the teeth earned the nickname Nutcracker Man.

The original cranium, when viewed from above, resembles a pitcher skull and is very extended in front thanks to the powerful development of the facial area, followed by a strong post-orbital constriction of the cranium just behind the eye sockets. Then the skull dramatically expands, huge zygomatic arches provide strong reinforcing elements, thanks to its helical curvature. The structure of the brain *A. boisei* was quite primitive.

Endocasts are characterized by strong narrow frontal lobes, even when compared with more ancient gracile australopithecines. *A. boisei* jaws were very large and massive. The relative sizes of the front and the molar teeth are very different from the proportions that are typical to gracile australopithecines, when the canines and incisors are reduced and the premolars and molars, in contrast, are hypertrophied. The enamel of the teeth is very thick; perhaps Australopithecus' food consisted of coarse fiber from tough vegetation. All the structure of the skull of *A. boisei* speaks of a big burden upon it; this is expressed in a large cranial wall thickness and hypertrophic jaws. Obviously, massive Australopithecus had adapted to feeding on plant food, and, apparently, the amount of food was very significant.

Along with the bones of *A. Boise* was found chipped pebbles and one can assume that they could use the stone for practical purposes or also it is possible that *A. boisei* simply fell victim to his contemporary Homo habilis who already could use stone tools. They evolved large teeth, powerful jaws and powerful chewing muscles. They themselves were very large, so that predators will not have found them easy prey. However, they did not use stone tools and their brain did not increase for 2 million years. They lived in Africa, near the ancestors of man, but human ancestors they were not.

Homo habilis 1960

In November 4, 1960 in Olduvai Gorge, Tanzania, Jonathan and Mary Leakey found the fossil, which was determined as specimen of *Homo habilis*. It consists of fragmented parts of a lower mandible with teeth, fingers, hand and wrist bones (Leakey, Tobias, & Napier, 1964). The remains are dated to approximately 1.75 Ma. The skull belonged to the child of 11-12 years old and judging by the structure of the foot, he was bipedal. Along with the bones of the new hominid were found remains of extinct saber-toothed tiger Smilodon, therefore, maybe he fell victim to the predator.

A new hominid was first named Prezinjanthropus, and then in 1964 was renamed *Homo habilis* at the suggestion of Louis Leakey. Leakey *et al* (1964) were confident of the belonging of this new hominid to the genus Homo. Later, similar findings were made in Koobi Fora, Swartkrans and elsewhere in Eastern and Southern Africa.

The height of *H. habilis* was around 1.3 m and its weight was about 50 kg; the face was an archaic form with supraorbital ridges, flat nose and protruding forward jaws. The volume of the brain was 600 cm^3 and the head became a more rounded shape than the typical Australopithecus.

More importantly, that *H. habilis* has a changed lobe structure; the more primitive occipital lobe of the brain is reduced in favour of enlarging the lobes of the more progressive frontal, parietal, and temporal lobes. The bulge in the thin-walled skull indicates the presence of their Broca's area (the centre of speech). The jaws were less massive than that of Australopithecus; the arms and thighs seem to be more modern; females have wider hips than males and this marked the sexual dimorphism of *H. habilis* (Sawyer, *et al.* 2007).

The structure of the hand had a mosaic pattern combining both progressive characteristics such as expansion of nail phalanges, indicating the formation of the finger pads as a tactile kinesthetic system and regressive characteristics: partial adaptation to climbing on trees. In addition, the hand had a strong grasp; no monkey has such abilities, which allowed producing stone tools. The *H. habilis* first toe was not allotted to the side and was located with the other fingers, so his foot was completely adapted to bipedal locomotion.

They intentionally made tools; the crudely worked stone pebbles of the Oldowan culture is often found along with the remains of *H. habilis*. Almost all their tools were made from quartz, which they had to bear between 3 and 15 km from a nearby quarry. One can say that they crossed the invisible boundary that separates the genus Homo from all other biological creatures. Classification of *H. habilis* as an independent species of the genus Homo caused broad discussion between anthropologists because many of the features of his anatomy were archaic bringing him closer to Australopithecus.

1974 Australopithecus afarensis

Australopithecus afarensis became well known from excavations at Hadar in 1973-1977 years, which were held by the International Afar Research Expedition under the leadership of Coppens, Taieb and Johanson (Johanson & Wong, 2009). The expedition found more than 240 hominid remains belonging to at least 35 individuals. Among the many findings, the most famous are the remains of the female AL 288-1 that has acquired the popular name 'Lucy' with the most probable dating of 3.2 Mya.

The species to which Lucy belonged was living in East Africa from about 4.0 to 2.9 Mya. There is a popular opinion that *A. afarensis* was almost certainly among human ancestors or at least was in a very close relationship with them. His primitive features such as the cranial capacity is between 375 cm^3 and 450 cm^3 combined with advanced structure of the pelvis and lower extremities that indicating bipedalism (Ibid.). The body size of *A. afarensis* is extremely varied. Possibly, the difference reflects sexual dimorphism, equal to that of gorillas and orangutans but greater than that of humans. The height of small animals barely exceeded a metre, while the larger ones were more than one and a half metres. The weight ranged from about 30 to 50 kg.

The endocast looks very similar to that of apes, but it does not have the lunate sulcus that is generally well marked on endocast of pongids and which modern man does not have. Parietal and temporal association areas increased, which certainly is a progressive sign. They had large incisors; their upper canines strongly protrude above the adjacent teeth and are larger in males than females.

The pelvis of *A. afarensis* is very wide and short and it is very different from the high-narrowed pelvis of apes and, in contrast, is little different from the pelvis of modern man. Apparently, specialization to bipedal walking among *A. afarensis* had gone very far. The morphology of the foot also shows the diversity of types of locomotion. In this respect, gracile Australopithecus was quite peculiar and different from both pongids and later hominids. In general, the hands were slightly elongated relative to the legs, and the legs were slightly shorter than that of modern man. These proportions are probably the result of incomplete transition to bipedal walking. Apparently, they walked slower and their daily distance was much smaller than that of modern man.

In this way, they are accurately described as bipedal apes with a monkey-like head. The earliest remains of Australopithecus found in Chad date back to 6 - 7 Mya, whereas the most recent dating of the finds has been determined for the massive Australopithecus in Swartkrans (South Africa) of 900 Tya. Thus, the period of existence of Australopithecus is extremely large. Within Africa, their location is concentrated in two main regions: East Africa (Tanzania, Kenya and Ethiopia) and South Africa. There are some findings of Australopithecus outside the continent, however in all cases are extremely fragmented and therefore controversial.

Homo rudolfensis 1978

In 1972, on the east side of Lake Turkana in Kenya, Bernard Ngeneo, a member of the expedition of Richard and Meave Leakey, discovered bones of a man which was taller and yet less massive than that of Australopithecus (Leakey R. E., 1973). The cranial capacity was measured as 750-775 cm^3 but later, in 2007, the construction of the skull of KNM-ER 1470 was downsized to about 526 cm^3. The relief and the wall thickness of the dome of the skull (calvarium) was less than australopithecines and the forehead is steeper than that of australopithecines.

The skull of KNM-ER 1470 was initially attributed to *H. habilis* because the remains were found in the same site and in the same strata as *H. habilis*. However, in 1978, Valery Alekseyev singled him out as a separate species *Homo Rudolfensis*. The unclear taxonomic status of *H. Rudolfensis* was initially caused by the complexities of dating the finding. KNM-ER 1470 was initially dated at nearly 2.87 Mya which places it as the predecessor of Homo but then the dating was corrected to 2.03 Mya, leading to the presumption of *H. Rudolfensis* being contemporary with *H. habilis*.

The chronological boundary of the existence of this species is quite vague. The gradual morphological transition between *H. habilis* and *H. rudolfensis* does not clearly define the moment of transition from one species to another. Moreover, gradual morphological transition between these two species (*H. habilis* and *H. Rudolfensis*) is a direct consequence of the continuity of evolution. Analysis of *H. Rudolfensis* shows the complexity of determining the systematic position for the paleontological findings when it comes to close species.

With a small brain volume, *H. rudolfensis* had large enough dimensions, approaching the size of modern man and a flat face allowed the association of *H. rudolfensis* with the older Kenyanthropus platyops that lived about 3 Mya (Leakey *et al.* 2001).

For around 40 years, KNM-ER 1470 was at the focus of the discussion over the number of species of early *Homo* present in the early Pleistocene. KNM-ER 1470 is a special case from other findings classified to early Homo because of its massive size and flat face. In 2012, Meave Leakey with her team discovered a face and two jawbones belonging to *H. Rudolfensis* with an age of between 1.78 and 1.95 Mya, making the anatomical and taxonomic status of KNM-ER 1470 less confused. The new finding KNM-ER 62000 is a well-preserved face of a late adolescent *H. Rudolfensis* that closely resembles KNM-ER 1470 but is remarkably smaller. The findings verify the existence of two contemporary species of early Homo, in addition to *Homo erectus,* between 2.5 and 1.5 Mya (Leakey M., 2012).

Homo ergaster (1984) Turkana Boy

In 1984, an expedition led by Richard Leakey managed to open an almost complete skeleton of a Homo erectus near Lake Turkana in northern Kenya. The find was truly unique because until that time, nobody could find such a well-preserved skeleton of early Homo. Further analysis showed that the skeleton belonged to a boy, presumably 12 years old, who lived about 1.54 Mya. It was named Turkana Boy and differed significantly with Australopithecus, having less powerful jaws and teeth. Most likely, he was a meat-eater and did not consume large amounts of coarse plant food and nuts in need of cracking; therefore, his jaw became more graceful in comparison with his predecessors (Leakey & Walker, 1985).

He had a very low forehead, which indirectly indicates that the frontal lobes of the brain responsible for complex thought processes were practically undeveloped. His parietal part of the skull, the centre of vision, was much broader than that of Australopithecus, which means that *H. ergaster* had much clearer vision than its predecessor. The cranial capacity was 880 cm³, that is a minimum level of *H. sapiens*. Broca's area of the brain suggests that he could have the beginnings of speech and that the set of sounds which he utilised could be quite varied. The skeleton had well-built brow ridges, a low inclined forehead and the absence of a chin.

The height of Turkana Boy was 1.6 m, as an adult he would have grown to more than 1.8 m, which shows a certain similarity of *H. erectus* to *H. sapiens* and the height of any kind of African Australopithecus were much lower. The pelvis of Turkana Boy was also much narrower than that of Australopithecus, which likely allowed him to move freely and faster on two legs. The life cycle of *H. ergaster* was not the same as modern humans; they quickly developed and were completely formed by the age of 12 years but they did not have sharp acceleration of growth that occurs in modern Homo teenagers during puberty.

The level of *H. ergaster* skill was significantly higher than that of all hominids known until that time. Whereas the early hominids relied mainly on flakes and large stones, relatives of Turkana Boy has used a relatively elegant stone axe and other processed stones that could lie comfortably in a palm. *H. ergaster* was the creators of the late pebble culture and they could probably use fire. Samples identified as *H. ergaster* have some clear resemblances with *H. habilis*. Consequently, many researchers consider that they belong to the same species. Whatever a researcher calls them, populations identified as *H. ergaster* are the likely ancestors of *H. erectus*.

Homo erectus georgicus 1991

In 1991, Georgian researcher David Lordkipanidze described the first remains of a new species, which were discovered at Dmanisi, Georgia. Stratigraphic and radiometric research and study of environmental fossil fauna shows an age of the findings of about 1.77 Mya. Thus, the Dmanisi finding was the oldest known species of the genus Homo that lived outside Africa. Lordkipanidze (Lordkipanidze, et al., 2007), described four skulls, on the basis of the analysis of findings it is expected that the height of Dmanisi hominoids was 145-166 cm and weight was 40-50 kg.

The skeletons of *Homo* Georgicus present a species with relatively advanced spine and lower limbs. The legs were like those of Homo erectus; apparently, they ran perfectly and could cover long distances but they were primitive in the skull and upper body, their hands and especially the structure of the shoulder joint were more like Australopithecus. A slight difference in the size of male and female also unites them with *H. erectus*.

Notably, the skull is from a male without teeth, in which almost all dental wells were overgrown with bone. The age of the deceased man was around forty years and the fact that the bone had grown into the dental holes indicates that he lived for a couple of years after his teeth fell out. Perhaps he was taken care of by his fellow tribesmen, which made it possible for a man who could not chew food to survive. Such care is an unexpected quality for those at such an early stage of evolution but the stone tools of *H. Georgicus* were relatively primitive and only a slightly better than Olduvai tools.

Study of the Georgian ancient hominids supports the popular hypothesis about that a small group of ancestors of modern humans probably migrated out of Africa to the Caucasus, where then they have evolved into *H. erectus*. Then they could go back to Africa, where they began their further evolution to *H. sapiens*. Another concept offers the idea that they were a descendant of the African *H. habilis* and an ancestor to Asian *H. erectus*. *Primarily, these discoveries were given the position of the species H. Georgicus, however subsequently it became the prevailing opinion that the Dmanisi findings are a local variation of H.erectus.*

Ardipithecus ramidus 1994

In 1994 in the Afar Rift region of northeastern Ethiopia were found teeth and jaw fragments, presented as *Ardipithecus ramidus,* a hominid species dated to 4.4 Mya. Later, 110 specimens were discovered and *Ardipithecus ramidus* bones were found in sedimentary deposits at a depth of approximately 3 metres located between two volcanic intercalations with an age of 4.4 million years. The biggest success was the discovery of a significant part of a well-preserved female skeleton, which weighed about 50 kg and stood about 120 cm tall. This find was unveiled to the general public under the name Ardi, whereas in official documents it is listed as Ardi skeleton ARA-VP-6/500 (White, *et al.* 2009).

The lower face of Ardi had a muzzle that protrudes less than a chimpanzee and her face is in a more vertical position than in chimpanzees. The cranial base is short from front to back, which indicates that the head was a top of the spine as in later bipedalism, instead of the front of the spinal column, as in quadrupedal apes (Ibid).

The teeth of *Ardipithecus ramidus* suggest that it was an omnivore and frugivorous with minor consumption of plant life and hard fibrous plant material. It is interesting that the size of the upper canine tooth was not distinctively different between males and females in contrast with the sexual dimorphism of chimpanzees, where males have considerably larger upper canine teeth than females (Suwa, et al., 2009).

This species had a small brain of around 350 cm³, much smaller than the brain of australopithecines at 500 cm³ and relatively similar to that of a modern bonobo. Ardi's skull is similar to skull of *Sahelanthropus. Ardipithecus* lived where dense forest interspersed with more rarefied woodland areas. They perhaps fed both in the trees and on the ground but consumed fewer open-environment resources in comparison with *Australopithecus.* The broad blades of the pelvis proves bipedalism of Ardipithecus. However, such features as the length of the hands, which reached her knees, and the ability to grab using the hallux (big toe) clearly show that these creatures spent a lot of time on tree branches (Ibid.).

Ardi's hands were preserved exceptionally well and they lacked specific features related to walking on knuckles like chimps and gorillas, a peculiar mode of locomotion characteristic only of the African apes. The hands of *Ardipithecus* are more flexible and agile than those of apes and a number of features are similar to humans and the hands of *Ardipithecus* allow her to walk on the branches leaning on her palms and they are better suited for using tools (Gibbons, 2007).

Until now, only two types of *Ardipithecus* were described: *Ramidus* and *Kadabba*. The latter was initially considered a subspecies of *Ardipithecus ramidus*, however the shape of the teeth, recently discovered in Ethiopia, identified a special specie. Kadabba lived between 5.8 and 5.2 Mya. The species name in the Afar language means 'common ancestor of the family' and it has been described as a probable chronospecies of *Ardipithecus ramidus*. The structure of the teeth of this species is similar to *Sahelanthropus* and *Orrorin*. Until 2009, it was assumed that A. Kadabba was the earliest known ancestor of chimpanzees and humans but now after a detailed study of Ardi, the separation of the lines of chimps and humans was shifted to 7 Mya (Gibbons, 2007).

It could be assumed that *Ardipithecus ramidus* is a perfect candidate for the role of a transitional link between the common ancestor of humans and chimpanzees and later hominids such as Australopithecus, from which came the first representatives of the genus *Homo*. The forests in which early hominids lived were not very dense and hominid bipedal walking was imperfect; a combination of a transitional environment with transitional gait supports the classical theory about the formation of bipedalism in conditions of cooling climate and shrinking tropical forests. One can say that hominids moved from dense forests to open spaces gradually and, as a result, gradually improved their gait.

Lovejoy challenges traditional ideas, where australopithecines learned to walk like a bipedal chimpanzee and he emphasizes that in reality chimpanzees and gorillas are distinct from other primates and are examples of relic primates surviving to the present day because they had taken refuge in the impenetrable tropical forests (Lovejoy, 2009).

One of the surprising conclusions that can be drawn from the study of *Ardipithecus* is that *H. sapiens* differs from the common ancestor in a variety of features, such as the size of the jaws and the anatomical structure of the hand and foot, less than chimpanzees differ from the same ancestor. If the level of progress is considered as the level of difference from the common ancestor then many human anatomical elements are more primitive than that of modern apes.

Comparison of *Ardipithecus* with *Sahelanthropus* once more shows that the evolution of human ancestors occurred with some step changes. Overall development in *Sahelanthropus* 6-7 Mya and *Ardipithecus* 4.4 Mya is practically the same, whereas after 200 thousand years in *A. Anamensis* one can see new features, which in turn changed little until the time of early Homo 2.3-2.6 Mya.

In the second half of the 20th century views on the phylogeny of hominids changed under the influence of numerous discoveries of different Australopithecus and Homo in Africa and Asia, as well as the appearance of new dating for previously discovered findings. The search for ancestors of Homo shifted to more ancient times and the difficulty of drawing the borders of taxa became apparent. It became clear that many lines of evolution of hominids, such as massive Australopithecus and Neanderthal are dead ends and the African hypothesis of the origin of modern man was formed.

Australopithecus Anamensis (1995)

In 1965, in the Kanapoi region of East Lake Turkana, a Harvard University expedition discovered fossils that consisted of an incomplete left humerus. It was about four million years old and was tentatively named Australopithecus (Johanson & Edgar, 1996). Later, a few other discoveries were made and in the 1990s, Meave Leakey and other affiliates organized a research team for the Kanapoi region. In 1994, they classified these hominid fossils as members of *Australopithecus afarensis* but in the following year, the specimens were reclassified *Australopithecus anamensis* (the word anam means lake in the Turkana language) due to some successful field seasons producing specimens with visible dissimilarity to the archetypal *A. Afarensis.*

All *A. anamensis* fossils were found inside a particular region, East of Lake Turkana, where there is a predominance of Pliocene sedimentary rocks that were deposited during an interval between 4.17 and 3.5 Mya, dated by radioisotopic methods (Leakey, *et al.* 1995).

A. anamensis had large canines, parallel rows of teeth and asymmetrical premolars and molars that were similar to those of apes. Fossils of the mandible, maxilla and a single temporal bone were found and these skull fragments carry numerous primitive, ape-like features. In addition, the fossils include the hindlimb, particularly the tibia, which show that this species may have walked on two legs. The knee-ends of the tibia are thickened and the tibial plateau, where the tibia connects to the femur is larger than in modern apes. They indicate that more weight was borne on the tibia, a feature essential for bipedality. Also, the tibia and humerus were more analogous to those from members of the genus Homo than they were to *A. afarensis.* The wrist bone suggests that this species had a limited ability to rotate the bones of the hand on the wrist, similar to later Australopithecus (Leakey et al. 1995). *A. anamensis* was slightly larger than *A. afarensis,* its weight was roughly 50 kilograms and sexual dimorphism was similar to that found in *A. afarensis.* Regrettably, there is insufficient material to estimate this hominid's cranial capacity and there is no evidence that they shaped stone tools or used fire.

According to environmental reconstructions, *A. anamensis* lived in forested habitats near streams and this disagrees robustly with the previously popular concept that bipedalism originally developed in open savanna environments. Rather, there exists the opposite opinion about the relationship of *A. anamensis* with other Australopithecus: some researchers suggest that it could be the direct ancestor of *A. afarensis* and that Homo is more closely related to *A. anamensis* than to *A. afarensis* (Johanson & Edgar, 1996; Ward, *et al.* 1999).

In 1996 a group of scientists led by Berhane Asfaw and Tim White in Afar Depression, Ethiopia found new remains which included parts of the arms and lower limbs of a few individuals and a partial scull. The remains were attributed to a new species and dated to roughly 2.5 million years ago. One year later, Haile-Selassie confirmed the new specimen and named it as *Australopithecus garhi*, where the word garhi means 'surprise' in the local Afar language.

A. garhi had an ordinary body dimension and had height 1.2-1.5 m. Its lower limbs were longer than other Australopithecus, demonstrating an ability for bipedal movement. However, their upper limbs were relatively robust and the same length as those of other Australopithecus. Its skull presents an exceptional combination of *A. afarensis* and *A. africanus* characteristics. The species unites a mainly primitive character of face and a highly unusual hominin dental configuration. *A. garhi* had massive post-canine teeth with big incisors and canines. Enamel thickness and occlusal form cannot align these teeth with *A. robustus* (Asfaw, 1999). The brain volume of *A. Garhi* was 450 cm^3, and cranial characteristics were comparable to those of *A. africanus*.

The peculiarity of this species lies in the fact that primitive stone tools were found along with the remains, resembling the Olduvai technology and bones of chopped antelope, which date from 2.5-2.6 Mya. This means that *A. garhi* started using stone tools before *H. habilis*. Some anthropologists put forward the idea that *A. garhi* represents a direct ancestor of modern Homo but others, judging by the dating and specialized features, suppose that it was not our direct ancestor (Wolpoff, 1999).

Reality of 21st century discoveries

Orrorin tugenensis 2000

In 2000, Senut and Pickford from the Muséum National d'Histoire Naturelle found and described the new genus of hominid in the Baringo district, Kenya. On the basis of bipedal locomotion together with dental and postcranial anatomy, the researchers deduced that the finding belongs to the hominid family. It lies between two strata of volcanic ash that were formed between 5.8 and 6.1 Mya during the Miocene. It received the generic name Orrorin Tugenensis meaning 'original man' in Tugen (Senut, *et al.* 2001).

The remains belonged to at least five different individuals. These include part of the femur, the spherical shape of which indirectly indicates bipedalism and a bone of the right hand adapted for climbing trees, but unsuitable for brachiation. Therefore, one can say that the species was previously adapted to habitual bipedalism, but that it was also a skilful climber.

Orrorin had small teeth relative to its body size and several teeth belonged to a species with a diet like Paranthropus. The molar size was closer to those of Ardipithecus and the molar enamel is thick. The anterior teeth, upper incisor and canine were closer in morphology to the teeth of female chimpanzees. The shape of the molars and canines is characteristic of animals that feed on fruits and occasionally meat. The size of Orrorin was comparable to modern chimpanzees.

The limb bones of Orrorin were larger than those of *A. afarensis*, signifying that the Homo ancestor may have been larger than previously imagined. O. Tugenensis' height was around 1.1-1.2 m or a little higher. The remains of animals and plants of the epoch when Orrorin existed show that they lived in dry evergreen forests but not in the savannah, as predicted by modern theories of human evolution. Therefore, the bipedal ape apparently first appeared in the woods and then later emerged into open space. Also, it is possible that walking upright was an element of adaptation to life in the trees.

The finding of Orrorin Tugenensis sets a controversial issue: if Orrorin is really a direct ancestor of humans, Ardipithecus and Australopithecus are a dead-end branch of human evolution as Orrorin possessed an appearance and a shape of femur much closer to human than Lucy. Furthermore, the finding supports the theory that the divergence between apes and humans probably occurred between 7 and 9 Mya (William & Haviland, 2010).

Kenyanthropus platyops 2001

In 1999, Meave Leakey's group at the semi-desert site on the western shore of Lake Turkana in northern Kenya found the 3.5 million-year-old shabby remains of a more or less whole skull and face of a completely new type of early human. It was named Kenyanthropus platyops that means the Flat-Faced Man of Kenya. The flat face was a big surprise for researchers as before the finding of *Kenyanthropus platyops* they supposed that hominins did not have a flat face until roughly 2 Mya. As noted by Leakey *et al* (2001), the facial architecture depends on the diet and how its jaws work, or more widely how these ancestors inhabited a diverse ecological niche. As a result, it is probable that Kenyanthropus platyops and Australopithecus afarensis existed contemporaneously, without direct antagonism for food resources.

Kenyanthropus had a mixture of archaic characteristics, such as a small ear canal, and more advanced ones like a flat face. The size of the skull and the theoretically calculated cranial capacity were similar to Australopithecus. The large cheekbones had an anterior position and the hard palate was wide. According to Leakey *et al* (2001), kenyanthropus was a direct ancestor of Homo sapiens but not Australopithecus afarensis that in some details was more modern than *Kenyanthropus platyops*.

Finding Kenyanthropus confirmed the idea of modern anthropology that modern humans did not follow a linear evolution in which one species evolved from the previous species. Instead, the discovery proposes that various species existed and each adapted to diverse environments (Lieberman, 2001).

Sahelanthropus tchadensis (2001)

In 2001, in the Sahel region of Chad, Michel Brunet and his colleagues discovered *Sahelanthropus tchadensis*, which was the oldest known human ancestor after the chimpanzee – human divergence. The team found an almost complete cranium and incomplete lower jaws. The age of the finding was established as between 6 and 7 million years old according to some associated fauna. The small cranium was named Toumai, which means in the local Chad's dialect 'hope of life' (Brunet, *et al.* 2002).

The volume of the Sahelanthropus brain was about 340-360 cm^3. The skull is elongated, which is more typical for monkeys. The occiput is flattened considerably compared to chimpanzees. The cranium has an orthognathic face with weak subnasal prognathism and a small braincase. It had several characteristics of later hominids such as small canines and intermediate postcanine enamel thickness; horizontal orientation of the basicranium, an anterior position of the foramen magnum, noticeably reduced subnasal prognathism and a big uninterrupted supraorbital torus, that indicate that Sahelanthropus belongs to the hominid clade (Ibid.).

On the other hand, it shows a collection of archaic features including a small brain size. Such variety of archaic and progressive features indicates its phylogenetic position as a hominid close to the last chimpanzee – human common ancestor and there is a possibility that this species was a sister group of more recent hominids. As the calculated age of Sahelanthropus is around 6 Mya, the divergence between the chimpanzee and human lineages took place before the advent of Sahelanthropus. However, some researchers (Wolpoff, *et al.* 2006) propose that *Sahelanthropus tchadensis* does not belong to human ancestry and connect it with the proto-gorilla. This would make *Sahelanthropus tchadensis* an ancestral relative of modern apes and so it could represent the first known member of their lineage. Furthermore, it indicates that the last common ancestor of humans and chimpanzees is unlikely to look like a modern chimpanzee.

In addition, this finding shows that hominids in the late Miocene occupied a relatively large geographical area and suggests that an exclusively East African origin of the hominid clade is incorrect. Analysis of the fossils collected along with the *Sahelanthropus tchadensis*, suggests they were living on the shore of a large lake around which lay savannah, fading into a sandy desert. The question of whether it is a bipedal creature remains debatable; advocates of this see evidence of bipedal locomotion of Sahelanthropus in the middle position of the foramen.

Homo floresiensis 2003

In 2003, on the Indonesian island of Flores, a joint Australian-Indonesian team of archaeologists unexpectedly discovered specimens of a nearly complete skeleton of a hominin (Aiello, 2010). Later, seven additional skeletons, dating from 38,000 to 13,000 years ago, were recovered.

The first partial skeleton, presumably an adult female, aged about 18 thousand years, which included a skull, jaw and most of the teeth as well as bone and teeth from seven additional skeletons, dating from 38,000 to 13,000 years ago, were found in the cave of Liang Bua in the west of the island, together with the bones of Stegodon and Komodo dragons. Moreover, together with the remains, stone tools were found that indicate the ability to hunt and build (Ibid.).

It was concluded that this was the discovery of a new species. Due to their height of around one metre, they received from researchers the nickname 'hobbits' and the species was named *Homo floresiensis*. The distinctive features of *H. floresiensis* are a small height and a small skull, lack of chin protrusion, flat nose and strong brow ridges. The theoretically calculated body weight was 16-29 kilograms, close to chimpanzees. A single found skull of *Homo floresiensis* had the cranial capacity of 417 cm³ that is comparable to Australopithecus; however, the brain-to-body mass ratio of the finding is close to *H. erectus.*

Some characteristics of the brain liken it to the brain of *Homo erectus*, while others such as the elongated occipital lobe of the cortex projecting above the cerebellum and the absence of fronto-orbital sulcus, which disappeared during evolution of the anthropoids as they developed frontal lobes of the cortex to the brain of *Homo sapiens* (Falk 2011). In addition, it has highly developed temporal lobes, relative to the total volume of the brain, which is involved in primary auditory perception and responsible for the identification of objects and persons. *H. floresiensis* had a relatively large Brodmann's area 10 of the dorsomedial prefrontal cortex, associated with higher cognition, despite the generally small brain size. This could indicate some degree of intelligence.

Despite its small height, the body of *H. floresiensis* was rather massive, similar to Australopithecus afarensis or small African apes. Characteristically, they had thick leg bones and a comparatively low twist of the arm bones.

It was found that Flores hominin arrived to the island not later than a million years ago, although the relationship of these ancient Homininae and *H. floresiensis* is not proven. There are a few sceptical hypotheses about the appearance of this type of Homo: representatives of one school see them as belonging to members of a dwarf species of Homo, which formed in conditions of island isolation. That is, the 'hobbits' may be a variant of habilis or Australopithecus, but this is not evidence of migration out of Africa, especially to Flores (Kubo, *et al.* 2013). Another interprets the remains as belonging to the disfigured pathology of modern Homo sapiens such as Down syndrome, myxoedematous and Laron-type dwarfism (Falk, 2011).

Kadanuumuu (2010)

n 2005, in the Rift Valley in the central Afar of Ethiopia, a new skeleton was found by a team led by Yohannes Haile-Selassie. The bones were located near the Mille River where the famous Lucy was discovered. It was named Kadanuumuu ('big man' in the Afar language), since the skeleton is estimated to be nearly 2 metres tall whereas Lucy was just over 1 metre tall (Dalton, 2010). The skull could not be found, but the bones of the left foot and right hand without feet and hands, a significant part of the pelvis, five ribs, a few vertebrae, the left clavicle and right scapula were found. Most likely, it was a male.

The skeleton gives an exclusive chance to assess some aspects of the locomotor habitus of *Australopithecus*. The Kadanuumuu morphology confirms a well-established terrestrial bipedality with only occasional use of the arboreal environment. Its pelvis and thorax were more similar to a modern human than ape-like and the scapula provides no evidence of a history of suspension or vertical climbing (Haile-Selassie *et al.* 2010). The ribs, pelvis and limb bones also demonstrate a variety of advanced features; the ratio of the length of the arms and legs fits into the range of normal variability of *H. sapiens*. Apparently, this means that the Afar australopithecines size and proportions of their body were quite variable.

Homo naledi 2013

In October 2013, in the Dinaledi Chamber that is located approximately 30 metres underground, in the Rising Star cave system (Gauteng Province, South Africa), fossil hominins were first recognized. During two relatively short field expeditions, in November 2013 and March 2014, the research team recovered an extensive collection of 1,550 hominin specimens, including 1,413 bone specimens and 137 isolated dental specimens, representing nearly every element of the skeleton multiple times, including many complete elements and morphologically informative fragments. The remains represent at least 15 hominin individuals from within a single population, with different body sizes and of both sexes. The large number of different sexes and ages of individuals makes this discovery the richest collection of related fossil hominins yet discovered in Africa. This previously unknown hominin species was named *Homo Naledi* where the word 'naledi' means 'star' in the Sotho language (Berger *et al.* 2015).

H. naledi exhibits an anatomical mosaic in which are presented features shared with Australopithecus, and with Homo, but also some features unknown in any hominin species. For example: the lower limb, the foot and ankle, the vertebrae the wrists, fingertips, and proportions of the fingers are largely Homo-like, but the pelvis, the shoulders the ribcage are like that of *A. afarensis* (Ibid). *H. naledi* had extremely elongated lower limbs, with mainly anthropomorphic ankles and feet, which propose better locomotor performance for running.

The skeleton shows that H. naledi stood upright and was bipedal. The teeth of *H. naledi* are smaller and markedly simpler in crown morphology than *H. habilis* and *H. erectus*. It has a range of body mass similar to modern small-bodied human populations, with results ranging between 39 kg and 56 kg. Generally, one can say that the anatomical structure of the species lies within the genus *Homo* rather than within the genus *Australopithecus*.

The brain size of *H. naledi* is similar to *Australopithecus,* approximately 560 cm^3 and 465 cm^3 for male and female respectively, however the morphology of the cranium, mandible, and dentition is predominantly coherent to *Homo.* Regardless of its small brain size, the cranium of *H. naledi* is structurally analogous to those of early *Homo*.

There have been different assessments of the findings. Berger suggests the findings represent a new species whereas other researchers are afraid to make estimates due to the absence of reliable dating of *H. naledi.* Some researchers declare that the fossils are comparable to examples of *H. erectus.* Furthermore, there is an opinion that the remains may possibly represent a relic population that could have developed in closed area of South Africa without relationship with other species (Kate, 2015).

In addition, some researchers hypothesize about the purposeful placement of the bodies due to some indications such as all remains belong to hominid, the absence of signs of predation or predator. There is no evidence of rock avalanches or a flow of water, which hypothetically could have entombed the bones within the cavern (Nadia, 2015). However, in this case we should consider ritual behaviour of ancient individuals, their ability for planned and conscious repetitive activity. However, this contradicts their cranial capacity, which is similar to that of *H. erectus*.

Conclusion

In general, over the last century, palaeoanthropology, from being a popular but rather exotic science built on the almost romantic enthusiasm of researchers, has become an industry that involves hundreds of expeditions all over the world and the number of finds is growing exponentially. The modern vision of anthropogenesis, accepted by the scientific community working within the traditional paradigm of evolution, is summarised below.

Around 10 Mya, some species of *Dryopithecus* changed their type of feeding, which resulted in a change in the configuration of the muzzle. Later, about 7 Mya, their descendants changed method of locomotion and consequently the morphology of the skull and post-cranial skeleton. This milestone is the moment of occurrence of *Sahelanthropus tchadensis* that may represent a common ancestor of humans and chimpanzees. Hypothetical constructions put *Sahelanthropus* as the main candidate for the ancestors of Ardipithecus, which was ancestor of *A. anamensis*. However, the discoverers of *Orrorin tugenensis* suppose that Sahelanthropus does not belong to the human lineage and it was a precursor of the gorilla. Moreover, the researchers assume that *Ardipithecus* and Australopithecus are a dead-end branch of evolution as *Orrorin* had a shape of femur much closer to human than to *A. afarensis*.

The first Australopithecus that evolved from mid-Miocene hominids lived in North and East Africa and later settled in the south of the continent, but South African Australopithecus were a dead-end branch of evolution. The major groups of gracile Australopithecus arose about 4 Mya and as discovered in Kenya, Tanzania and Ethiopia. Fossils of *A. afarensis* originated from about 4 to 2.5 Mya and probably, they were a direct predecessor of the genus Homo. *A. afarensis* were upright-walking creatures of about 1-1.5 metres tall.

Apparently, Australopithecus walked with shorter steps than modern Homo and their hip joints were not completely unbent when walking; their hands were somewhat elongated and the fingers were adapted for climbing on trees. These features can only be inherited from ancient ancestors but they had quite a modern structure of the legs and pelvis. The latest gracile Australopithecus is found in Ethiopia *A. Garhi* dating from about 2.5 Mya. Primitive tools were found along with the remains, similar to the tools of Olduvai. This shows that *A. Garhi* started using stone tools before *H. habilis*.

Bipedal, robust Australopithecus and Paranthropus emerged in East Africa from unknown early australopithecines around 2.5 Mya. They had extreme hypertrophy of the masticatory apparatus. Moreover, the later individuals, with prognathism of the face and a large brain, were relatively similar to early Homo. Gracile australopithecines, with a primitive structure of the skull and postcranial skeleton and a small brain, continued to exist for a very long time simultaneously with the occurrence of robust australopithecines of South Africa that emerged from the local gracile forms 2.7 Mya but these South African australopithecines are an evolutionary dead-end.

East African gracile Australopithecus are most likely the ancestors of hominids. Between 2.5 and 2.3 Mya they changed their lifestyle by becoming omnivorous. As a result, the masticatory apparatus decreased, the skull sharply gracilized, creating conditions for the significant increase of brain and consequently the way opened to the emergence of the family Hominidae.

The first East Africa hominids converged with Paranthropus increasing the masticatory apparatus and highly progressive morphology of the braincase. The close descendants of these early hominids were characterized by high growth and they were among the first hominids to exist outside of Africa.

Hominid migration in Europe and Asia began about 1.5 Mya. Migration between Europe and Asia took place in several stages. Anthropological studies show that about 780-600 Tya Asian hominids were almost identical to those from Europe. However, about 400 Tya when the flow from Europe and Africa ceased, Asian Homo had specific features. Around 120 Tya in Europe and Africa, there was a new type of Homo that was closer to H. sapiens in the structure of the cerebral box and the lower jaw. However, it had specialized features in the structure of the facial part of the skull and the postcranial skeleton. Cro-Magnon man appeared about 40 Tya. They were no different from a modern man in appearance and physical development. The Cro-Magnon people appeared later than the Neanderthals and coexisted with them for some time. Indeed, interbreeding between Cro-Magnon and Neanderthals played a part in the emergence of modern man. A completely modern morphological complex was formed only in the region of 25 thousand years ago.

Chapter 2

The analysis of the simian approach

Despite the astonishing development of palaeoanthropology, there is little empirical understanding of the critical issues involved, namely: how an almost naked creature so poorly physically adapted to the wild environment could survive whilst leaving a comparatively safe tier of the tropical forest and walking on the ground on short, crooked and weak legs. Nearly one hundred years have passed since the publication of Scheler's 'The Human Place in the Cosmos', but his critical question is still relevant. How did man evolve from his fourhanded ancestors?

A metaphysical focus on simian theory

Palaeoanthropology has unearthed a huge number of artefacts and, through the use of unique technology, has been able to synthesize the approaches of different sciences. The discovery of new fossils over the last five decades and the development of new methods for the studying and dating of fossils, as well as reconstructing their locomotion and paleoenvironment have significantly increased knowledge about the morphology and evolutionary relationships of anthropoids. One of the main goals of palaeoanthropology is to find the ancestor of humans from among the common stock of hominids that constitute a group of African and Eurasian forms that lived during the Miocene epoch and are considered to include the common ancestor of humans and apes.

An analysis of current data of anthropogenesis has allowed many scholars to conclude that science has solved the problem of the relationship between man and apes in line with Charles Darwin's simian hypothesis. Indeed, with the advent and further development of Darwin's evolutionary theory of the origin of species and humans, supporters of the simian theory have concluded that all other hypotheses are weak.

On the issue of human origins, the dominant theory over the last 100 years has been the most conventional simian theory of anthropogenesis, whose scientific status is established. According to its followers, all other theories are unworthy of attention as they do not have sufficient support in modern biology. Simianists accept that a long time ago human ancestors were not yet people but apes. Therefore, it is necessary to establish the point at which an ancient ape became man. The problem with identifying the point of this transition was codified by Charles Darwin, who recognized the impossibility of identifying a specific point at which an ape-like creature evolved into a man.

With the accumulation of paleoanthropological artefacts, the distinction between primitive humans and apes became blurred. The find of Pithecanthropus by Dubois on Java was a composition of two different individuals - human and gibbon. It should also be noted that the new find of *Australopithecus Sediba*, described in 2010, was closer to apes because it had a small brain, long arms and relatively short legs and shoulders like a modern orangutan. However, its pelvis, thighs, ankles and spine were closer to human. In the structure of the hand, ape and human features are mixed (Berger, et al., 2010). As a result of the discovery, Berger's article proposed that the finding might be a transitional species between Australopithecus and *Homo habilis*.

In 1948, Sir Arthur Keith introduced the concept of a cerebral Rubicon between ape and man. It is reasonably thought that the most important difference between man and ape is not in the structure of the teeth, nor the composition of the body, but in the brain. Keith compared the biggest brain of the gorilla (650 cm^3) and the smallest brain of modern man (about 855 cm^3) and decided that the point of transition should be between these values. According to Keith, any group of apes whose brain reached 750 cm^3 should be considered sapiens, not anthropoids.

Keith managed to check his criterion on findings known at the time such as Javanese Pithecanthropus and African australopithecines. Pithecanthropus' average brain volume was 850 cm^3, and its structures were dominated by human features. Australopithecus, having had a monkey-like skull, had a brain volume of less than 650 cm^3. This corresponded to the concept of Keith but after a few years, anthropologists found fossils of people with smaller brain capacity, such as 'the Hobbit' from the island of Flores which had only 426 cm^3 and the concept crumbled. The concept of the cerebral Rubicon gradually dried up until it lost scientific significance. Numerous attempts to find the morphological criteria for distinguishing between these forms were in vain. This cerebral Rubicon was refuted in practice when it turned out that there is not a simple correlation between the size of the brain and mental ability.

Currently, the cerebral Rubicon belongs entirely to the history of science and is an example of the vulgarization of the scientific approach. According to general opinion, the anatomical criterion is impossible in principle. The border is now considered in cultural rather than anatomical terms; the estimate of the degree of development of humanity of a hominid is possible based on the existence of tools together with remains. At the same time, the triad of attributes of humankind continues to be present in the theory, not as a kind of positive knowledge about the course of evolution, but as an implicit methodological approach.

In general, if one takes the whole array of accumulated data and arranges it in chronological order, one can see how in ancient primates there appears, progressively human traits over time as the monkey characteristics fade and give way to human ones, until finally, a moment comes when one can tentatively say it is a present-day man.

One can plot the known findings on a time axis to show progress in the development from an ancient stratum to a younger one. It can be seen especially well by the example of those places where various hominids were found such as Olduvai or Omo, where these findings cover large time intervals. In all cases, one can see that after Australopithecus came *H. ergaster*, then *H. erectus* and afterwards came *H. sapiens*. Archaeology shows a similar picture; there are not tools in layers with Australopithecus, primitive pebble tools are found in layers where early Homo is found and above it is the more evolved symmetrical axe.

Evolutionists note that human evolution is not a simple and linear process; in the evolutionary paths were bizarre twists, branches and deadlocks. Some of the human populations scattered across the planet became stuck in their development and some degraded. Since the formation of humans was a long-term process but evolution is continuous, between great apes that lived 10 million years ago and modern man there were a lot of species and it is possible to talk about many missing links. Intermediate links are found between already intermediate links and finds are so numerous that their multiplicity becomes a problem for anthropologists. Meanwhile, new kinds of fossil hominids continue to be uncovered with true regularity. With each discovery, with every newly described species, the picture becomes more and more problematic. It turns out that a lot of bipedal Hominidae creatures have lived on Earth. For example, between 2 and 4 Mya in Africa, there were several species of Australopithecus. Therefore, one can be sure that in the near future currently unknown new species will be described.

According to the simianist postulates, one of Australopithecus gave rise to the human species but also there is no doubt that they cannot all be simultaneously our ancestors. They assume that there could be hybrids formed of several primate species. Therefore, simianists raised the question of the trigger mechanism of anthropogenesis which may be formulated as why some species of hominoids managed to start using their intellectual and cultural potential much more actively than others, having had approximately the same abilities, anatomical and physiological features of the nervous system.

The development of bipedal locomotion

A variety of hypotheses explains the appearance of people in the centre of events that were a key to start of the process of hominization. Primarily, the occurrence of bipedal locomotion, a large brain and a working hand, so-called morphological triad of characteristics of humanity and resettlement of hominids from the tropical forests to savannah. The author of the triad idea was Engels (1876), who in 'Dialectics of Nature' developed the idea of the triad and drew a hypothetical scheme of the formation, in a positive feedback process of growing complexity of labour, of the three main anatomical and physiological characteristics of man: working hands, an upright posture and a large brain.

Engels suggested the existence of a complex pendulum-like motion from biological to social regularities and back during anthroposociogenesis. According to the Engels theory, due to changes in natural conditions, human ancestors increasingly used the products of nature (stones, sticks and bones) in their activities. This led to the development of simple work skills, which in turn were the impetus for adapted hands. The making and using of tools led to an additional development of the hand, consequently that the hand was simultaneously both the *product* and the *organ* of labour. This made it possible to create more complex tools and led to the division of labour, which is a collective process in which people have a need to communicate. As a result, the undeveloped larynx of the ape slowly but inexorably transformed and the mouth gradually began to pronounce one articulate sound after another. This in turn stimulated the development of the brain. As Engels (1954, p. 59) noted 'First comes labour, after it and alongside it, articulate speech. These were the two most essential stimuli under the influence of which the brain of the ape gradually changed into that of man.'

The main weakness of this hypothesis is that it is impossible to establish a causal link between the appearance of man and that of labour activities. In order that 'labour created man,' labour had to have existed before Homo, namely apes had to work, that was not possible due to their skills and absence of causes that forced to them to work.

Engels pointed out that the upright posture and bipedalism liberated the hands for the manufacture of tools and was a crucial step in the transition from ape to man, therefore one can begin analysis of the anatomical and physiological characteristics with bipedalism because the evolution of bipedal locomotion is a significant problem in the lore of human origins.

This problem raises questions of why bipedalism was selected for; how, when, and why it developed and why it became such a successful form of locomotion for Homo. In addition, there has been a significant scale of relatively polarized discussion with understanding of bipedalism. However, regardless of the scientific outlook of researchers, transition to an upright posture is, for many of them, the trigger of the start of the formation of the human morphological complex.

Under the generalizing concept 'bipedalism of man', one can mean a complex biomechanical phenomenon consisting of three separate elements: orthograde stability, bipedia and vertical locomotion. Each of these may be found in the animal world only in isolation and does not require any explanation in the framework of anthropogenesis. For example, meerkats can stand for the purpose of orientation, but they move on all four limbs, that is it orthograde stability without locomotion.

Orthograde stability, which is not uncommon in nature, is possible for many rodents (squirrels, marmots, rats and mice), monkeys and bears.

Dynamic bipedia was typical for many species of reptiles in the Jurassic period. In modern times, it is common in certain lizards and is a characteristic feature of birds. Among mammals, dynamic bipedia is characteristic of the kangaroo that moves by leaps on hind limbs, but the vertical standing on their hind legs is absent; at rest, they are based on three points, including a powerful tail and their hind limbs in contact with the ground at the level of the knee joint. This dynamic bipedia without a stable orthograde is very different from human walking. In both cases, it is not at all similar to what man can do; the human way of movement is unique in its complexity.

Upright walking of man only occurs in the simultaneous presence of the three elements of bipedal locomotion. However, no one element taken separately can be considered a purely human trait. Moreover, any other mammal moving on two legs, such as a monkey, bear or kangaroo, does so with a fairly pronounced forward tilt, whereas for humans a strict vertical position is an identifying characteristic. In a certain sense, it is an external criterion of the species if one considers the importance that people give to a vertical posture.

There are many different hypotheses for the origin of bipedal locomotion and the degree of consensus and clarity is inversely proportional to the growing number of hypotheses. In theoretical constructs postulating one or another sequence of interrelated events in human evolution, this item is a weak link, one that could completely collapse the theoretical construct.

Most of the early theories about why people became bipedal emphasize the freeing of the hands as the main force of selection. Darwin in 'Descent of Man' (1871) wrote that man could not achieve his current dominant position in the world without using the hands, which are so delightfully adapted to serve. But while hands were regularly used while moving, they could hardly be sufficiently committed to making weapons, or the aiming of stones and spears. Consequently, it was beneficial for man to become bipedal.

From Darwin's time, numerous researchers offered a large quantity of different theoretical models of bipedal locomotion and generally, all noted that bipedal locomotion is an extremely exceptional form of primate locomotion. In this circumstance, Lovejoy (1981) notes that bipedalism could develop only because of the strong pressure of natural selection and for a long time skeletal alterations related with bipedalism were substantial.

Most authors raising the question of the origin of bipedal hominids believe that, from the beginning, this property gave some advantage to its owners; otherwise, it would simply not arise. In line with the common hypothesis, the transition of human predecessors to the upright posture was explained by the necessity of adaptation to the life in the savannah because primarily forested environments were gradually replaced by more mosaic environments made up of different degrees of open grassland and open woodland (Reed, 1997).

To cope with these environmental changes hominids had to adopt a series of new behavioural strategies. Changes in habitat composition would have resulted in a shift in food availability and thus necessitated a shift in food acquisition behaviours (Rose, 1991). Hominoids had either to range further to find food or develop strategies to procure other, new types of feeding. Consequently, it is possible that hominoids would have had to have engaged in more terrestrial travel over more open habitats, and it is in this change in activity patterns that one is likely to find the reasons for the emergence of bipedalism.

Relocation from the forest to the savannah could be associated with any cooling, increase or decrease in the dryness of forest cover in the habitats of great apes. For instance, Van Couvering (2000) notes that modern theories have tended to relate to paleoenvironmental changes from the end of the Miocene through to the beginning of the Pleistocene. The beginning of habitual bipedalism strongly correlates with generally cooler and dryer global conditions, and an associated increase in more open grassland habitats. The date of the alleged relocation varies according to different authors, from seven hundred thousand years ago to 6-7 Mya (Ibid.).

Observation of the habits of mongooses, marmots and ground squirrels that live in open spaces and stand upright for the purpose of orientation has led to an interesting hypothesis that hominids straightened because of the need to see far and better navigate in the savannah. In these environments, good observation over the top of long grass is needed for the search of prey and for the timely detection of predators. However, this theory contradicts the fact that to some extent bipedal Ardipithecus and Australopithecines inhabited tropical rain forest landscapes and so the transition to two legs was not, and could not be, related to the adaptation to the open landscape (Mithen, 2006).

Approximately of the same heuristic ability for the explanation of human bipedalism is the hypothesis that the upright man has a psychological advantage over the four-legged animals, as he looks down and seems larger because of the working principle of those above being the stronger animals in the world. Alternatively, bipedalism occurred because males had to bear food to the females with whom they mated.

Washburn (1960, 1967), developing the ideas of Darwin and Engels, suggests that bipedalism was needed for the hands-free motion necessary for tool-carrying behaviour and gave man a number of important advantages over other animals. However, modern archaeological data show that the first stone tools appeared two million years later than the first bipedal hominids (Rose 1991).

The bioenergetic hypothesis explains the emergence of bipedalism by greater energetic efficiency of the human compared to the quadrupedal walking of apes (Rodman & McHenry, 1980). However, this theory is appealing to advantages related to bipedal locomotion, which may occur in a fully formed bipedal human, but which are almost absent in the process of its formation in the early stages of Homo evolution. Moreover, there is not strong evidence that ape locomotion is less physiologically efficient than that of modern Homo.

There is a hypothesis linking becoming two-legged and the liberation of the forelimbs from locomotor function with the need to carry food and cubs and to scare away predators by throwing stones. However, this hypothesis contradicts observations of modern gorillas that are capable of carrying cubs and scaring away predators by throwing stones using their forelimbs as hands, but remain quadrupedal.

Hunt (1996) examines chimpanzees when they sporadically use bipedal locomotion to pick fruit. Chimpanzees employ their hands for balance and manage to move gradually from tree to tree to collect fruit. Hunt supposes that bipedal movement provides power consumption savings due to the bipedal position avoiding raising and then lowering their body weight. He supposes that Lucy's anatomy was more suited to standing than walking and the arched fingers and strong upper limbs should be considered as adaptations for hanging on one arm at the same time as standing on two legs and feeding with the other arm but not for climbing trees.

There are also more complex multi-factor theories, which are a compilation of simple ones. Robinson (1972) notes that there are many reasons for the appearance of bipedalism. Probably, there was a combination of selective factors robustly relating to feeding and reproductive behaviour, which changed locomotor performance in the hominid. Walking on two legs provided the ability to range further for food and water, giving hominids the opportunity to explore new territory and new food sources. As a result, this formed new behavioural strategies to manage such variation.

Supporters of the thermoregulatory hypothesis have seen a reason for the transition of human predecessors to become bipeds in the fact that an upright posture makes available distinctive thermoregulatory advantages in open savannah habitats because bipeds would receive considerably less direct solar radiation when standing erect (Wheeler, 1984, 1992). Wheeler estimates that, at midday, a hominid in upright position would absorb 60 per cent less heat. Their bodies, being located higher up away from the land, would be cooled down by convective air currents. These factors would significantly decrease the rate of solar insulation of hominids on open ground, allowing them to move further without water intake. However, this raises a counter-question: why would they climb down from the trees, in the shade of which the problem of excessive insulation does not exist? On the contrary, it would be better to go into the forest. Chimpanzees live in the second tier of the rainforest, where the soil has almost zero exposure to sunlight.

Another popular concept is that hominids became walked upright long before they had to leave the jungle and began to live in the savannah. Discoveries made in recent years have shown that early hominids, along with still imperfect bipedalism, retained some skeletal features such as the curvature of the phalanges of fingers and toes, considerable mobility of the upper limb at the shoulder joint and several other features related to arboreal life. This theory suggests that there was a period when the ancestor of man was not perfectly mobile on two legs, as he became later, yet he had already lost some skill in climbing trees.

Keith (1948) expressed the idea that the basic anatomical features that provide movement on the ground without forelimbs were formed in the process of adaptation to the tree-based lifestyle rather more than a terrestrial lifestyle. That is, walking upright would be a kind of inheritance received from preceding stages of evolutionary history. Therefore, the original type of forest locomotion caused the formation of the anatomical prerequisites of bipedal hominids, requiring a vertical position of the body and lower limb support functions such as vertical climbing or brachiation. As a result, specialization to a similar way of movement developed sufficiently strongly in the transition to a terrestrial existence, so even imperfect bipedalism was more convenient for them than four-legged motion; the consequence of which was the preservation of this feature in the new environment.

All modern apes and fossil of hominids such as Australopithecus and Paranthropus (with the exception of humans) do not have the foot's two longitudinal and one anterior transverse arches; that is, they have flat feet. The human foot is a complex damping mechanism containing 26 bones, 33 joints and more than a hundred muscles, tendons, and ligaments. The arches contribute to the overall spring action of the legs and feet and they make bipedalism possible. Without these spring actions, human walking and running would be much more laborious. Of course, an arched foot can pass as flat, but the reverse is not possible. Unfortunately, in the days of Lucy's primeval forest, there were no orthopaedists. Therefore, a question is justified: how did flatfooted monkeys walking on two legs not get arthritis of the talonavicular foot joint, but rather a spring-arched foot?

Thus, it turns out that in the framework of the classical simian theory, to find some specific benefits that could be associated with bipedalism in the early stages of its development is practically impossible, so compelling reasons to transition to orthograde locomotion have not yet been found. Now we should consider the question of what made human ancestors change from their arboreal lifestyle to the ground.

Why monkeys climbed down from the tree

Human evolution has been closely linked to environmental and climate changes. Due to the cooling and draining of the climate in Africa during the Miocene and Pliocene, the area occupied by tropical forests gradually decreased, and the jungles that prevailed previously were replaced by savannah. The climate became increasingly unstable; the temperature and the amount of precipitation significantly reduced and fluctuations of temperature and humidity between seasons increased. Climate and landscape change affected the speciation of hominids. Environmental change forced the hominids to master new ecological niches and the search for such niches inevitably led them to the ground. In particular, around 3 Mya *A. Afarensis* disappeared, and it was replaced by gracile and robust Australopithecus and early representatives of Homo.

According to the simian theory, a trigger mechanism of anthropogenesis can be represented as follows. When under pressure of environmental factors in the Pliocene, hominid ancestors in East Africa moved to a predominantly terrestrial life because their anatomy was formed so that bipedalism became for them a more optimal mode of locomotion. However, bipedalism was not a significant evolutionary advantage over other animals. Something was needed which could compensate for this imperfect bipedalism. This something was culture. Having actively adapted to the natural environment by creating artificial environments, hominids first overcame the cultural Rubicon which determined the direction of their future evolution. The significant difference of the evolution of hominids from other animals was the fact that they had to adapt not only to the natural environment but also to a cultural one, which itself also constantly developed. As a result, anthropogenesis led to the appearance of the many behavioural and anatomical features that distinguish humans from other apes.

However, it may be noted that during the past few million years, the zone of tropical forests where the apes lived always remained quite extensive so that apes were preserved. Anthropoids did not have the urgent need to drastically change habitat; therefore, the role of geographical factors in shaping human evolution cannot be accepted as essential (Foley, 1987).

There is the idea of nutrition of human ancestors by leaves, twigs and fallen fruit that could be acquired easily being below branches. Such forage, which is energy-poor, required special adaptations of hominoids to break into this ecological niche. Animals living in this environment are forced to consume food in large quantities, chewing almost without interruption. Therefore, they have huge, powerful jaws, encompassing most of the skull, with an extended abdomen that can contain a huge amount of rough forage. However, the large jaws and simple lifestyle meant that the brain could not develop; because of this, over million years of this lifestyle, a potential predecessor could evolve into the gorilla, who feeds under the branches, dwelling in the lower tier of the rainforest.

It is known that some animals give birth to many children, but at the same time care for offspring is almost absent and they have high mortality. Others give birth to fewer, but devote fully to childcare and selflessly protect them, for example, female chimpanzees give birth every three years and keep in touch with their children throughout their lives.

Knowing this, Lovejoy (1988) offers that the reason for the transition to the bipedal locomotion of apes was the change in their breeding strategies. They evolved to give birth more, while on the forest floor it is easier to take care of several calves than was the case on the second tier. The question arises: what prevented apes giving birth more frequently in the trees, or rather, on the second tier of the rainforest, where even soil is available? Why would they come down to earth, teeming with predators? Sitting at the top, apes constitute uneasy prey for relatively few predators, whereas on the ground, they are an easy target for anyone who has strong fangs and hunger. Johanson & Shreeve (1989) note that the idea of Lovejoy is excellent but lacks a trigger mechanism of bipedalism. The transition to the new breeding strategies related to the annual births has nothing to do with the biped. There is no correlation between these two phenomena. In trees, many mammals give birth annually rather than once in three years, as do chimpanzees, but they do not move to the ground. This lack of evidence for a trigger mechanism is a general deficiency in all hypotheses of the transition to upright posture. The transition to an upright posture placed predecessors of Homo, for example, Australopithecus, upon the brink of disaster, and it is reasonable to ask how did bipedalism occur and why it was permitted by natural selection?

Since it was for them a disadvantage, both in relation to predators and in relation to the competitive monkey types, then other things being equal it would make their chances of survival in a crisis situation very low. In this case, according to simianists, it needed a culture that could compensate for underdeveloped locomotion. However, it is not clear that if monkeys by nature are capable of complex intellectual activity, why they needed a cultural revolution in order to stand on two feet. Almost by divine miracle, the transition to an upright posture and culture, of animals that did not have the need and opportunity to make this transition, puts the evolution of man in the category of accidental events (Lindblad, 1987).

Trying to explain the transition to upright posture, Lindblad suggested imagining yourself as a monkey that had to evolve in order to survive the transformation from a peaceful inhabitant of the forest, as a mainly frugivorous primate to a carnivorous predator that lives in such a dramatically different environment as the sun-scorched savannah. By virtue of common sense, a monkey should maintain a protective scalp and in order to pursue swift inhabitants of the savannah it should remain faithful to the style of running on four legs. Limbs had to be equal or nearly equal in length. Instead, the extremely long arms and short legs of forest monkeys underwent the opposite change; humans are hairless, bipedal primates that have long legs and short arms. For killing prey long, sharp fangs were required, as those of the leopard, not the relatively small stumps of the hominids. The successful product of mutation, rather, would be a creature like a baboon, with huge teeth and a long nose (instead of the more flat snout), with a thick coat of hair, limbs of almost equal length and a more horizontal posture (Ibid.).

Failure to clearly explain the origin of bipedalism is not the only drawback of the simian theory of anthropogenesis. The morphological approach has revealed the discrepancy between the time of formation of the main attributes of the hominid triad and inclusion of the labour factor in the evolution of hominids that was taken as the cause of these attributes. In addition, according to archaeology, there is no connection between the beginning of the stone industry and the development of intellectual abilities. The nature of the use of stone tools by *Homo erectus* indicates rather an instinctive work than a rational one with an absence of progress for millions of years and, as a result, labour is not identified as a causal factor of progressive evolution. On the contrary, the fact that there is a clear advantage of speed, quality and quantity of simple labour operations from narrow specialization of labour starting as an instinctive reflex activity suggests that the first human ancestors' labour operations led these human ancestors to a dead end of narrow specialization.

The transition from instinctive labour to conscious labour is an integral part of the more general problem of the origin of consciousness. It became clear that it could not appear gradually; a solution to the problem of the transition from instinctive labour to conscious labour depends on solving the general problem of the origin of consciousness, rather than vice versa.

A sufficiently long domestication experiment, carried out over thousands of years in some species, proves that neither labour nor educational exercises specifically aimed at developing the rudiments of pet minds, give results which could be interpreted as the continuous emergence of the mind. At the same time, we know that in the process of domestication and artificial selection such morphological changes occur very rapidly. In fact, Darwin was able to present natural selection as an experimental confirmation, delivered by humanity during domestication. However, in the case of the occurrence of the mind, the application of the evolutionary mechanism described in the monograph 'On the Origin of Species' was not sufficient to describe human development in the book 'The Descent of Man'.

At the beginning of labour activity lies no obvious trigger mechanism. As already mentioned, so far nobody has been able to explain convincingly why simian ancestors of humans needed labour as the main driver of evolution. When many researchers say that a decisive factor in human evolution was the emergence of instruments of labour, they often forget to mention that many animals have sophisticated tools. For example, the spider web is simultaneously a complex instrument of labour and an object of labour. Anthropologists have not yet been able to empirically prove the idea of Engels. However, recognition of labour as a factor of human evolution is heuristically justified.

Ten (2011) suggests human labour as a causal factor of the evolution of *H. sapiens*, who could express themselves only in the form of art, and immediately became creative. That means, as a minimum, the connection of two abstract concepts. For example, predators taking a stone to split a skull of an animal to get the brain is not labour as it fits into the stimulus-response scheme because a direct link can be traced between the two objects. Many animals are capable of such actions, such as sea otters, which crack the shells of sea urchins on the rocks. But to tie a stone to a stick or give it a supplement in the form of some kind of strong thread that immediately multiplies its destructive power is creativity, this is the way to develop the mind. Here, the connection of two objects abstract from a direct task and a third object, which is necessary to influence. This is subject of rudimentary abstract thinking. Hence, at the beginning of human labour were not Olduvai pebbles but some two-part tool. Stones used by *Australopithecus* and *Pithecanthropus* have a random character and therefore do not relate to progressive evolution.

The introduction of absolute dating methods has revealed a time difference of millions of years in the emergence of the signs of the hominid triad (working hand, bipedalism and a large enough brain). There is a clear lag after the emergence and improvement of bipedal locomotion but prior to brain development. Over the past 30 years, archaeologists have found bipedal hominids (Sahelanthropus, Orrorin, Kenyanthropus platyops) dating 6.5-3.5 Mya and brachiators with an age of less than 2 Mya that violate the harmony of the theory. The idea that the structure of the front limb is directly related to the adaptation to a sustainable upright posture, or conversely to brachiation, came under question as the criterion of working hands. For example, Homo habilis, who lived about 2 Mya, is connected to the first archaeological Olduvai culture and had such features as a curvature and relatively large length of the phalanges of the hands. Expansion of the adjacent ends of the metatarsal bones testifies in favour of brachiation (Danilova, 1965). As a result, Homo habilis did not develop in the hominid but rather in the pongid direction. In view of this, Australopithecus cannot be considered as the first man.

Recently, several hundred different hominids were discovered, many of which were presented as a scientific sensation. As a result, these hundreds of sensational finds were able to convince the community of the correctness of the simianist theory. Nevertheless, a huge amount of paleoanthropological research that uses the most modern methods of science nevertheless does not give clear answers for easy questions such as who was the common ancestor of hominids and pongids from which began the divergence?

One of the highlights of criticism of the simian theory is the lack of progenitors of anthropoids. There is an evolutionary time gap between the tailless apes and tailed primates. The anatomy of anthropoids is specific, their intelligence is several times higher, and no transitional links are evident. If man evolved from some ancient pongids, it is necessary to discover a common ancestor from which two branches divided.

If human ancestors are to be found in the person of the famous African and Asian hominid, the ancestors of chimpanzees has been found by no one. Johansson and Idi (1984) write that modern gorillas, orangutans and chimpanzees seem to appear from nowhere. Today they are with us, but they do not have a yesterday. Fossil *Dryopithecus* disappeared around 8.5 Mya and the intermediate species between them and modern anthropoids are unknown. Nevertheless, researchers sometimes announce findings of the ancestors of apes. For example, Kelley (2002) believes that finds in the Indian subcontinent *Sivapithecus* dated from 2.2 Mya may have been the ancestor of modern orangutans.

Generally, primatology exists in a strange situation, which does not have a logical explanation. Primates have existed for tens of millions of years and they are very well studied, except in one important aspect: the ancestors of apes among primates cannot be identified in the fossil record. After all, this is a most intriguing question because we are talking about the ancestors of beings from which people supposedly descended. Studying tailed monkeys, both modern and fossil, did not give an answer to the question of the line leading to the tailless Hominoid apes. Who were they - monkey of the savannah or of the forest? Were they like baboons, monkeys or macaques? It is not possible to trace the genetic link and the inexplicably changed morphological structure.

Scientists are unable to give an evolutionary response to even such a seemingly simple question as why apes have lost a tail, which is not a redundant organ in the tropics where there are many mosquitoes, if monkeys have it and widely use it in everyday life. Why was it necessary to lose the tail? How did it come to be absent in orangutans and gorillas? Conversely, the tail for arboreal monkeys is like a fifth paw, it is very convenient to hang by a tail on a branch so two forepaws are free. It could not reduce, as the reduction of an organ occurs when the organ remains without utility and it atrophies. Rather, the tail is constantly used, as for instance in the Howler monkey.

The many paleoanthropological discoveries do not give the highly anticipated answers to the most insistent mysteries such as which hominid line leads to Homo sapiens. What was the trigger mechanism of anthropogenesis? How and why there was a transition to upright posture? How did consciousness originate? The theory of anthropogenesis came to the point where the study of particulars does not give new knowledge because there is a large gap between discoveries in palaeoanthropology and their interpretations.

The source of consciousness as a key problem

If the problem of the origin of bipedalism is so entangled in anthropogenesis, the problem of the origin of consciousness can rightly be placed on the opposite pole: it is a virgin topic in palaeoanthropology. The question of the origin of consciousness is closely connected with the question of the essence of consciousness and, as such, is not the exclusive theme of palaeoanthropology. Moreover, it seems that a significant proportion of paleoanthropologists consider the central theme of this science to be the formation of complex human morphology, and problems associated with the formation of the psyche, are investigated insofar as they are included in anatomical investigation. For example, are or are there not Wernicke's and Broca's areas? How did frontal lobes develop? This purely materialistic approach, for all its importance, does not give any suggestion about the idea of intellectual-mnemonic ability.

Summarising the wide variety of modern anthropological hypotheses about the origin of consciousness in the chronological pattern of encephalization, Ruff *et al.* (1997) note a period of increase between 2 and 1.5 Mya. After that was a million years of stagnation, the cause of which is not clear, followed by a more impressive growth of brain size between 0.6 and 0.25 Mya which is most probably linked to the evolution of language.

According to the simian approach, the expansion of the brain can be explained by new requirements of sensory-motor control that arise as a necessary element when walking upright. Standing or walking on two legs requires that balance is constantly monitored, and small muscle groups are included in the work to maintain the position of the body; the movement of the legs should be coordinated with the arms and body to preserve the dynamic equilibrium (Aiello 1996).

Hunting, carrion eating and activity with tools demanded planning, communication and hand-eye coordination that also expanded the brain. As soon as the hands were released from a locomotory role, they could be used for new activity such as carrying, signalling and making tools. Theoretically, the picking of fruit made available the selective forces for vertical locomotion that, in turn, freed the hands for making stone tools and further evolution (Mithen 2007).

The increased consumption of meat in the diet led to a decrease in the size of the intestine and is apparently linked to the evolution of bipedalism as well as brain size. That provided the cognitive abilities to formulate more complex foraging strategies and production of stone tools, which in turn allowed a further increase in consumption of meat and other quality foods such as fruit (Mithen 2007). Hominid bipedalism allowed the exploitation of hunting niches unavailable to other carnivores. Bipedalism made it possible to increase the area covered as well as the foraging time and freed hands for the production, processing and use of stone tools (Isaac 1989).

According to Dunbar (1988), primate group size is bigger in open environments than in wooded environments because in savannah only a large group can deter predators and reduce the likelihood that any one individual will be preyed upon. For this reason, the increasing aridity in the Pleistocene decreased forest cover and probably enlarged hominid group size. As a result, large social groups could create a more powerful need for communication than those arising from foraging ranges and production of tools. Communication, in turn, created the preconditions for increasing the brain. Dunbar (2004) notes that the increase in brain size of Homo genus indicates increasing group size. This idea is supported by archaeological evidence suggesting that the earliest groups of Homo were larger than those of *Australopithecus* and *Paranthropus*.

The consequences of this for cognitive and cultural evolution were significant because the group provided better defence against predators, better opportunities for feeding and information transmission that creates the preconditions for the further development of the brain. Increasing complexity of social interaction in large groups may reflect increased intelligence that in turn increased brains of Homo. Moreover, some persons could forecast the behaviour of others due to better cognitive abilities and consequently they received an advantage within the tribe. Thereby it is possible to talk about the beginning of social selection.

Many modern paleoanthropological approaches to solving the problem of the origin of human consciousness do not look convincing despite an abundance of findings, a variety of scientific methods and a large number of theoretical constructs. Most approaches are based on a behavioural paradigm, which is characterized by the denial of the gap between the mentality of apes and human consciousness. Trying to reduce the gap between man and animal, researchers identify animals by human characteristics, thereby trying to show that there is no special problem of the origin of human consciousness.

Using again and again in their theoretical constructs such words as possibly, probably, little by little, gradually and step by step the researchers turn their theories into fascinating home reading, however it does not answer the key questions of anthropogenesis.

Conclusion

According to the most conventional simianist approach, the main event of human evolution was the transition of the Hominoid apes from a predominantly arboreal forest to predominantly terrestrial existence in the open or mosaic landscapes and development of bipedal locomotion. It is believed that hominoid transition to life in the open landscapes gave them a need to adapt to an unfamiliar environment, pushed them to seek new ecological niches and stimulated the developmental activity of tools and sociality. The development of bipedal locomotion, which had the result of the liberation of the forelimbs from locomotory function, is a premise of this development.

The main problem of anthropogenesis cannot be considered as solved, even in general terms, because it cannot explain what led to the change of environment, nor what resulted in changing the way of movement. Most importantly, it does not explain why these two events were not made in the usual adaptive biological way, or what stimulated the creation of culture and intellect.

Continuing research efforts concentrate on three main areas: the working hand, the brain and bipedalism. The neglect of many other exclusive anatomical and physiological characteristics of man leads to the impossibility of constructing a harmonious scientific theory.

Chapter 3

Aquatic Ape Theory

The Aquatic Ape Theory (AAT) is a theory of human origin, which in contrast to the generally accepted theory based on the savannah dominant factor that made apes move to bipedalism, considers the shallows of seas, rivers and lakes as the dominant factor of human evolution. The AAT supposes that going through an aquatic environment acted as an agent of selection in the evolution of humans more than it has in the evolution of modern apes. Consequently, many of the most important physical dissimilarities between human beings and apes may be better explained as adaptations to wading, swimming and diving. Ordinary scientific observation pushed the researchers to the idea that the ancestors of man could have lived in an aquatic environment, which formed the specific water characteristics of modern man.

The unique path of man

According to the AAT, approximately 6-7 Mya some group of human predecessors were separated in a water environment of East Africa where the large part of the territory was flooded that created conditions for the existence of the aquatic ape. Accordingly, they lived in a water environment and managed to seek out nutrition in the water, and after some time, their anatomy changed to be more successful in the water. Being isolated, this group of human predecessors could have much faster evolution where mutations in isolated populations significantly accelerated the evolutionary development. Consequently, these populations of animals lost hair and became more streamlined, whilst retaining fat. Having fat under the skin made them more buoyant in water. The theory is intended to explain the difference between man and other anthropoids by the aquatic pathway of existence of their ancestors. One of the first to draw attention to seemingly minor human characteristics that distinguish humans from the ape community was Wood (1929) who noted the smooth hair (lanugo) that can be observed even in human embryos and newborn babies and subcutaneous fat that is dramatically thicker than that of monkeys.

Developing the ideas of Wood, marine biologist Hardy (1960) put forward the hypothesis that human ancestors were not wood or savannah monkeys, but water (sea) monkeys. He proposed that: *'a branch of the primitive ape-stock was forced by competition from life in the trees to feed on the sea-shores and to hunt for food, shellfish, sea-urchins etc., in the shallow waters off the coast. I suppose that they were forced into the water just as we have seen happen in so many other groups of terrestrial animals. I am imagining this happening in the warmer parts of the world, in the tropical seas.'* (Hardy, 1960, p. 642).

Hardy paid attention to the incredible perspiration of man that could help predators to find them easily in forest or savannah. For this reason, they needed regular contact with water. In addition, he noted that early man could walk upright only in the water, where his body weight under the influence of buoyancy force is much smaller. The human type of hair growth on the head, Hardy explained as due to aquatic animals being more susceptible to sunstroke than land animals. This is questionable, given that aquatic animals have the ability to constantly wet the head, thereby cooling it. In addition, he drew attention to the ability of people to dive under water for a long time, while apes are deprived of this ability.

He singled out different human exclusives such as a large nose, having compared a human nose with that of Indonesian proboscis (long-nosed) monkeys, which conduct a semi-aquatic lifestyle in the mangroves. However, he could not explain the connection between a large nose and a water lifestyle. The nose of man and the nose of the aquatic monkey differ significantly from each other not only in size but also in anatomical structure. The long nose of a man is a hydrostatic device for diving. The nostrils of man constituted in the likeness of a deepening bell; while diving man is unconsciously straining nose, extended in the external part and narrowed in the internal, the nostrils form an airlock. Water cannot overcome it, so men can dive. This very versatile anatomy allows people to have both water and land lifestyles. The noses of the proboscis monkeys are an elongated cavernous monkey nose. These monkeys are descendants of land monkeys who have overcome the characteristic for all monkeys of hydrophobia. Unlike those of humans, their noses were not formed in water but on land, then simply lengthened to gigantic proportions as a morphological adaptation of monkeys to a new environment. This is a narrow specialization, an evolutionary dead end. With a dangling nose longer than ten centimetres, a monkey is doomed to lead only an aquatic life in mangroves and is unable to return to the ground, into the rainforest, since in the battles with other monkeys over the territory this nose would become a weakness. With the transition of monkeys to the aquatic mode of life, the task was to provide an obstacle to the flow of water into the brain. Therefore, evolution went along the path of lengthening the nose. Thanks to this, the monkey manages to come out of the water before the water flows through the labyrinth of the long nasal cavity.

Even though that the AAT was independently created by Hardy in 1929 he managed to publish it only in 1960. Therefore, the first complete and printed AAT was presented by the German pathologist Max Westenhöfer in 1942. During the 1930s and 1940s, he published several publications where he analysed the relationship between *Homo sapiens* and the apes. He assumed that Homo could not be in close relation with primates, because Homo has less specific characteristics, which are nearer to the original mammalian design than arboreal primates. As a result, he proposed that the early relatives of human may have been amphibian. He argued that many modern human characteristics derived from a completely aquatic way of life in the open water, and that humans only in recent times returned to land; these arguments were some of the first descriptions of AAT. He noted that the postulation of an aquatic mode of life during an early stage of human evolution is a reasonable theory, for which additional investigation may uncover extra supporting proof (ibid.)

Westenhöfer highlighted a number of facts found by different researchers, supporting the AAT: '*examination of water mammals shows that the antihelix, tragus and antitragus and their corresponding muscles have retained the ability to close the ear canal when diving in water and the human ear also must have had this ability at some point in its development. There were persons whose ears' structure comes closest to that of the original, closable form and these muscles are found to still be contractible. Also, physiological experiments show that electrical stimulation of the human tragus and antitragus muscles can close the entrance of the ear.*' (Westenhöfer, 1942, p. 148-150).

In addition, he underlined that the human spleen and kidneys seem much more similar to those of whales, dolphins, seals, sea otters and bears than to those of apes. This link between such dissimilar species can be explained as a convergence that is the process of evolution of distantly related groups of organisms, the acquisition of a similar structure as a result of their existence in similar conditions and identical direction of natural selection. Dolphins, seals and man, in comparison with the macrosomatic animals, have a less developed sensory system used for the sense of smell due to a particular development of the brain. Therefore, these data confirm the idea that in a period of human development there were predecessors of man with an aquatic lifestyle.

The human foot that widens in the direction of the toes theoretically indicates a marshy habitat as per the feet of some fossilised bog animals. The comparative smoothness of accessible seafood would mean that robust maxillary apparatus would become redundant. He paid attention to the fact that a man has a deficit of hair that is similar to the relative absence of hair in water mammals and that the subcutaneous fat of humans has capacity for expansion. He noted that in the skin of modern man's back there are atavistic mucous glands that are similar to hippopotamus glands. In addition, comparatively conventional web-like skin formations on the hand and toes are as in the semi-aquatic giant otter shrew; the body hair direction in the direction of the elbow on the lower arm in Homo and apes would have been useful when advance arms forward while swimming. According to Westenhöfer, all these anatomical features are examples of a physiological adaptation to the aquatic environment that is supplementary confirmation of a preceding aquatic path of man. He noted that the form and development of the Adam's apple, the nasal cartilage, the ear muscle, the shortness of the outer auditory passage, some characteristics of the female genitals and the development of the kidneys might establish verification of a possible aquatic phase in human evolution (Ibid.). Regardless of Westenhöfer theory being the foundation of the AAT, numerous AAT supporters note that his concept of the aquatic theory has several inconsistencies and primitive details.

Morgan's wetland ape

In the 1970s, Morgan continued the scientific establishment of the aquatic ape theory of human evolution. Moreover, her book The Descent of Woman (1972) became a bestseller, turning her into a feminist star. She noted that many morphological and physiological characteristics of man, which were typically considered as exclusive to man, in reality, are not really unique at all. They may often be common, in some cases almost universal, in groups of mammals, which have left the land and returned to an aquatic environment. Again and again, across evolution this development occurred. Different animals from reptiles to mammals that were completely terrestrial animals set aside their earthly way of life, plunged into the water, and adapted in various ways.

She noted that there are different types of aquatic animals such as penguins, sea lions, aquatic insectivores (water shrew), aquatic hoofed mammals (hippopotamus), aquatic marsupials (water opossum), aquatic rodents (beaver) and aquatic reptiles (crocodile). These existing mammalian species took to the water and then developed exclusive adaptations for an aquatic existence. Consequently, Morgan supposed that some the species of apes could take a similar aquatic pathway and they could be direct human ancestors.

Morgan supposed that about 7 Mya an area of low-lying land, Afar Triple Junction, which is part of the Great Rift Valley in East Africa, was flooded by the sea and became the Sea of Afar. As a result of this environment change, part of the ape population there needed to replace their territory, moved to off-shore islands or lived in flooded forests, salt marshes, mangrove swamps, and lagoons. Some of them survived and managed to adapt to the aquatic environment. Later, when the Sea of Afar evaporated, their descendants began to migrate toward the south, following the waterways of the Rift Valley.

There is an ecological argument for the existence of a wetland ape; wetland habitats could have been populated by apes leading to their adaptation to a more water-based lifestyle. Ellis (1986, 1987) assumes that when hominins left woodland into aquatic areas they had to experience an environment with a smaller amount of competition and more safety from predators. Moreover, wetland habitats had high-energy foods, which give additional opportunities for large brain growth. The change of feeding behaviour from a foliage diet to meat was compulsory for savannah apes because on the grasslands animal-food was under the control of more powerful animals and therefore its availability would have been limited. In contrast, tropical marine environments could provide plentiful seafood all year round and gathering it did not require skill, entailed small risk with no serious competition and, as a result, the inclusion of animal food into the diet for human predecessors would be very much easier via the marine food chain (Ellis 1986, 1987).

The AAT postulates a period when considerable areas of North Africa were overwhelmed by the sea, except a few mountainous areas that formed islands with the shallow marine areas and then later the level of sea declined. Consequently, the islands once again became part of the continent. One group of apes took a divergent course of evolution through adaptation to a water environment in the same way as other species had done previously. However, when the water receded the adapted water apes received new ecological opportunities and managed to come back to terrestrial existence bringing new features of inherent aquatic adaptations, which considerably affected their evolution (Morris, 1967).

It is not too difficult to show that Homo sapiens are similar to apes and they share a common ancestor but it is more complicated to give an explanation of why Homo differs from the gorilla and the chimpanzee much more notably than they differ from each another. There is conventional understanding that Homo ancestors carried on along a completely dissimilar evolutionary path, in comparison with apes, because a Homo predecessor moved out of the forests onto the open savannah. But in this case one should see distinctly human features, adapted to a savannah environment, and some of these adaptations should be paralleled in other savannah mammals. The AAT suggests that when Homo ancestors moved onto the savannah they already had particular human features such as nakedness and bipedalism that they had begun to develop much earlier, when the monkey and human lines first diverged (Morgan, 1997).

Verhaegen *et al.* (2002) hypothesise that gorilla and pan could move away from inhabiting coastal swamp to the inside of Africa up the rivers and later, when the continent became arid, they started adaptation to overland life and a quadrupedal locomotion. Homo, meanwhile, inhabited the coasts but finally began to dwell in a few forested coastal areas. Possibly, Homo populations appeared as coastal omnivores consuming marine food sources and migrated along the coasts of the Indian Ocean and Mediterranean Sea.

Aquatic features of human anatomy

The theory stresses that many of features of human anatomy and physiology, though relatively unique among terrestrial mammals, are common in aquatic environments. Therefore, most of the outstanding issues of human evolution become much easier to unravel if researchers accept the idea that the earliest ancestors of Homo found themselves living for a prolonged period in a flooded, semi-aquatic habitat. Morgan offered powerful evidence to support this hypothesis and some of this evidence is briefly outlined below.

The naked ape

There are hundreds of existing primate species, however only humans are naked. Moreover, all other naked mammals, which have lost practically all of their fur, such as dolphins, walruses, manatees, hippopotamuses, pigs and elephants, exist in water rich environments like marshes, lakes or the sea. Traditional anthropology proposes that Homo became hairless under the sun in the African savannah but this contradicts zoological observations where the fur of different mammals (camel and fennec fox) acts as a defence against the heat of the sun. Whereas fur makes the best insulation for terrestrial mammals the best insulation in water is a layer of fat.

Fat

Biologists note that humans have dramatically more fat cells in their bodies than other apes and more than would be expected in an animal of human size. In addition, there are two different types of fatty tissue: visceral fat and subcutaneous fat. In aquatic mammals, and in humans, a higher proportion of subcutaneous fat is deposited just under the skin. Visceral fat, on the other hand, is stored around the internal organs in abdominal area, including the liver, pancreas and intestines. In land mammals fat tends to be stored internally, especially around the kidneys and intestines; In hibernating mammals the fat is seasonal; in most aquatic ones, as in humans it is present all the year round. Morgan noted that there is little chance that early man would have developed subcutaneous fat following moving to savannah and becoming a hunter, because it would have slowed him down. Human fat is an inheritance from an earlier aquatic phase of human evolution. Also, all newborn primates except modern Homo are slim because their ability to cling to their mothers and support their whole weight with their fingers is critically important for survival. Human offspring accumulate fat in the prenatal period and for several months after birth. Moreover, much of this adipose tissue is white fat that is good for insulation in water and for giving buoyancy and exceptionally atypical in newborn mammals in comparison with brown adipose tissue that is better for supplying energy.

However, as noted by Pond (1987), this type of adipocyte is found in humans, fin whales, hedgehogs, monkeys, and badgers. She also points out that the human fat distribution, in contrast to that of whales and seals, could be an element of a non-aquatic adaptation and the most important function of fat is as an energy supply rather than insulation as for arctic species. This point contradicts Morgan's idea that fat was an adaptation for insulation in an aquatic environment and buoyancy in a water-side dwelling.

Walking on two legs

Morgan offered an explanation of human bipedalism as the human environment began to be flooded, so they would have been affected to move on their hindquarters when they came down to the ground in order to keep their heads above water. Her supporters believe that wading through shallow water was a major factor of the locomotion of the earliest proto-hominids that took to habitats where considerable wading locomotion was favourable; the differences between the postcranial anatomy of the earliest bipedal hominids and modern humans could be explained by particular adaptations for wading.

Breathing

There are differences between the respiratory systems of humans and terrestrial mammals. The first is that human can consciously control their breathing whereas other mammals commonly cannot and so make these actions involuntary in the same way as the processes of digestion. Only aquatic mammals have breaths with voluntary control when they decide how deep they are going to dive can estimate how much air they need to inhale. Human voluntary breath control gives an opportunity to speak. The second difference between humans and terrestrial mammals is the human larynx which a few months after birth descends into the throat, right down below the back of the tongue. As a result, humans can respire through their mouths as well as their noses. Morgan supposed that is an example of aquatic adaptation, because aquatic animals such as penguins, pelicans, gannets, the sea lion and the dugong as well as man, when swimming, require gulping to quickly inhale more air through the mouth than through the nostrils. A terrestrial mammal is usually has to breathe constantly through the nose, because its trachea proceeds through the back of the throat and the top end of the larynx is located in the posterior of its nasal passageway.

System of perspiration

The human system of perspiration, in comparison with other mammals, is very wasteful of the body's essential resources of water and salt due to its use of different skin glands. From an evolutionary point of view, it is doubtful that humans acquired these skin glands on the hot savannah, where there is a deficit of both water and salt.

There is a small number of species such as the walrus, the seal and the sea otter, which have ability to weep tears, which may be triggered by an emotional excitement caused by feeding, fighting or frustration. However, there is no other land animal except human, which can weep tears.

Also, every square centimetre of human skin has from a few to hundreds of microscopic sebaceous glands which exude sebum, an oily secretion, over the head and body. Consisting of a lipid blend, sebum increases the antimicrobial barrier of the skin, lubricates hair and the epidermis and gives the skin elasticity. The greatest production is reached during puberty under the influence of testosterone in males and progesterone in females. Sebaceous glands of chimpanzee have rudimentary character. Other animals use sebum only for waterproofing of the skin or the fur.

Nutrition

It is known that aquatic mammals more frequently develop large brains whereas mainly terrestrial mammals, on the contrary, stagnate in brain development. In connection with this, Morgan suggested that the human brain encephalization was a response to increased consumption of seafood: development of a human brain is maintained by different micronutrients required throughout life in small quantities to arrange a variety of physiological functions such as iodine and omega-3 fatty acid. She believed that contrast between human and ape brains could be partially determined by nutrition due to the building of brain tissue being dependent on a sufficient provision of essential fatty acids, which are comparatively limited in the savannah food chain but yielded in great quantities by the marine food chain. In addition, Cunnane (2005) supposes that increased consumption of molluscs, clams and oysters that gave docosahexaenoic acid (DHA) and iodine by pre-Homo human ancestors could have encouraged the enlargement of their brain size.

Numerous fossils of *Homo georgicus*, *Homo erectus* and *Neanderthal* demonstrate extensive tooth loss, bone resorption of tooth alveoli and generalised enamel dysplasia existing from birth. Such dentitional atrophy in comparison with australopithecines' large cheek teeth and apes' large canine teeth is mismatched with partial meat consumption on savannas (Fischman, 2005; Zilberman, 2004). However, it can be easy explained by getting some benefit in a shore-based lifestyle where consumption of soft shellfish might have been important. This also explains the abrupt reduction of human masticatory musculature that, according to supporters of the savannah theory of evolution, does not support the evolution of man from the savannah where it was necessary to have the sharp dentition of dogs and hyenas (Stedman *et al.* 2004; Verhaegen *et al.* 2007).

Other features

The aforementioned animals also have outstanding olfactory abilities to detect corpses, while homo does not have similar abilities because a sense of smell was not significant for the collection and consumption of shellfish by human ancestors. This is similar to numerous underwater foragers that have undeveloped olfactory lobes in the brain compared to terrestrial mammals. Waterside mammals that forage crabs and shellfish, for instance racoons, marsh mongooses and cape clawless otters have unwebbed and highly sensitive and mobile fingers. Homo, and to some extent other primates, also have extra feeling and more mobile hands than most other terrestrial mammals. This, in some ways, reveals an aquatic path of human origins (Verhaegen *et al.* 2007).

Over the course of some five million years, Homo readapted to terrestrial life and therefore one can anticipate that the footprints of aquatic adaptation have become moderately obscured. Nevertheless, traces still exist and Morgan selected human characteristics, which are present in aquatic animals and practically absent in apes and savannah mammals: sebaceous glands, ventro-ventral copulation, hymen, loss of vibrissae, volitional breath control, eccrine thermoregulation and descended larynx, eccrine thermoregulation, diminution of apocrine glands, enlarged sebaceous glands and emotion-triggered tears.

Criticism of the Aquatic Ape Theory

Despite the fact that there are a number of arguments supporting the AAT, they are not adequately influential to neutralize the arguments against it. Most modern anthropologists do not support the position of Morgan. For example, Moore (2003), in summarizing the position of a large number of critics notes that even though the AAT followers say that an aquatic lifestyle generates opportunities for bipedalism, even they should agree there is no strong evidence for this statement. They also exclude numerous terrestrial mammals, which make use of sporadic bipedalism. He remarks that there are no aquatic mammals that employ occasional bipedalism for the same reasons humans do: locomotion, feeding, guarding and display behaviour. However, Moore disregards Hardy's creative suggestion that wading in shallow water could encourage bipedalism of human predecessors. Hardy wrote *'Wading about, at first paddling... He would naturally have to return to the beach to sleep and to get water to drink; actually I imagine him to have spent at least half of his time on the land.'* (Hardy, 1960, p. 644).

In addition, critics of the Aquatic Ape Theory's simplified bipedalism of man note that it is a complex bio-mechanical phenomenon consisting of three separate elements: orthograde stability, bipedia and vertical locomotion. These critics analyse one separate element that may be easy found in the animal world only in isolation. For example, squirrels, marmots and monkeys are able to stand for the purpose of orientation, but they move on all four limbs, that is orthograde stability without locomotion. An example of dynamic bipedia is the kangaroo that moves by leaping on their hind limbs, but do not stand vertically on their hind legs; at rest, they are supported by a powerful tail as well as their hind limbs. The upright walking of man occurs only in the simultaneous presence of all three elements of bipedal locomotion (more details of this problem are disclosed in Chapter 4).

Moore (2003) contends that the Morgan hypothesis of loss of human body hair is highly misleading because only large and fast-swimming cetaceans and sirenians that have lost their hair and specialised for that niche over tens of millions of years became aquatic animals. However, AAT supporters postulate that when an individual finds an abnormal characteristic in a species that is unexplainable, one looks for analogies in other species. As noted by Morris (1967, p. 16) describing Homo sapiens as the naked ape, 'staring at this strange specimen and puzzling over the significance of its unique features, the zoologist now has to start making comparisons. Where else is nudity at a premium? The other primates are no help, so it means looking further afield. A rapid survey of the whole range of the living mammals soon proves that they are remarkably attached to their protective, furry covering, and that very few of around 5,000 living mammal species in existence have seen fit to abandon it. At this point the zoologist is forced to the conclusion that either he is dealing with a burrowing or an aquatic mammal, or there is something very odd, indeed unique, about the evolutionary history of the naked ape.' (Ibid. p.17)

Kuliukas (2016) remarks that there are eleven possible evolutionary examples of nakedness in the mammalian order. Whereas five may be attributable to aquatic forces, two may be attributable to large ranging animals like elephants, two for burrowing such as the naked mole rat and the last two are pigs and humans. If we suppose that Homo ancestors were never as large as the elephant and did not have a subterranean existence, aquatic factors are moderately likely evidence as to what may be behind human nakedness. Therefore, nudity can be a premium as homo was moving through water.

Opponents of Hardy note that human fat deposits are similar in pattern and life history to other primates and dissimilar to fatty aquatic mammals. However, this thesis is destroyed by the strong arguments of Montagna (1985), who notes that simultaneously with the lack of fur; skin acquired a hypodermal fatty layer that is significantly thicker than exists in other primates. In addition, Montagna says that analysis of the ratio of apocrine and eccrine glands between different species of apes shows that man has more eccrine sweat glands and fewer apocrine glands on his body surface and this could be a modern aspect of evolution.

Not all primates are inclined to store subcutaneous fat in located in specific depots around the body. The subcutaneous human fat is an extension of the typical fat depots and is positioned in locations that differ from aquatic mammals' subcutaneous fat that is dispersed in a thick layer over the whole body (Pond 1991). Pond proposes that the sex difference in fat deposition, where women accumulate body fat around the buttocks, thighs and on the breasts can be partially explained by sexual selection. That is, selection would support selective mating by males with women with these extra reserves.

Additionally, Reynolds (1991) notes that a subcutaneously dispersed deposit of fat should raise body temperature. However, hominids on the savannahs of Africa would encounter cold night-time temperatures and subcutaneous fat helped adaptation to these varied conditions. During the daytime the bare skin and sweat adaptation complex does extremely well whereas at night subcutaneous fat is much more useful as an insulator. Consequently, the thickness of the fat layer was optimised to keep the body temperature warm enough at night without obstructing heat loss during the day. In general, the arguments for an aquatic origin for human body fat look fragile, as the savannah arguments seem more influential (Ibid.).

Sweating is a particularly efficient adaptation when there is a bountiful provision of water and the air is relatively dry. Australopithecus, Homo habilis and other early hominids lived in a lakeshore environment, where drinking water was not a scarce resource and the climate of East Africa was mainly dry and hot. In such conditions, sweating can be understood as a basic adaptation to a savannah environment rather than an aquatic one (Ibid.).

Both humans and various aquatic animal use the larynx, which lies in the throat rather than the nasal cavity, to close off the trachea while diving and for facilitating large breaths of air upon surfacing. Moore (2003) suggests that there is a difference between the human larynx, together with that of aquatic mammals, and the more descended larynx discovered in a range of terrestrial animals such as chimps and deer. However, aquatic mammals such as dolphins have been fully aquatic for tens of millions of years and therefore the evolution of a blowhole took the larynx in an exceptionally distinct way. Only the walrus, sea lion and dugong, that are less aquatic animals, have such a similar characteristic (Kuliukas 2014) that supports Morgan's theory.

Convergence in evolution exists when unrelated animals share comparable peculiarities that occur due to a similarity in function of these peculiarities. The consequence of convergent evolution is a convergent similarity. That is, the similarity of organisms, not based on affinity but on a closed feature set, formed independently in different groups. The main cause of convergent evolution is considered to be similar ecological niches of these organisms. An example of convergent evolution is the formation of a body of similar form in predatory sharks and dolphins.

The ecological niche of large mobile aquatic predators is the same for all similar groups and puts forward similar demands upon the shape of the animal's body. But evolutionary convergences have their restrictions: eventually, differences will emerge. Dolphins and sharks are similar in form, but these similarities do not make sharks into dolphins. Therefore, every trait should be examined as an individual functional and evolutionary issue. Many critics of AAT note that their approach employs excessive environmental determinism. That is, the postulate that evolutionary changes are constantly driven mainly by the environment. Morgan (2008) and other supporters of the aquatic theory argue that the concentration of the modern world population demonstrates the intention of people to repose in coastal regions alongside seashores, lakes and river valleys. However, their opponents note that humans are weak swimmers in comparison with aquatic animals. Most terrestrial mammals have a cutaneous muscle (*panniculus carnosus)* layer located throughout the skin. Animals employ twitching muscle to get rid of insects and dirt. However, humans, unlike other primates are missing the *panniculus carnosus.* Possibly, human predecessors could jump in the water to keep away from pests.

It is important to note the existence of the dive reflex that appears when animals are submerged in water, their heart rate slows down and the blood supply is redistributed from the skin to organs for maintaining oxygenation. Aquatic animals have a stronger dive reflex than terrestrial animals (Blizniak, 2012).

In addition, there is a difference in blood composition between water and terrestrial animals, as well as between humans and apes. The size of red blood cells (erythrocytes) varies considerably, for example in humans its diameter is 7-8 microns whereas goat cells are half the size. In general, erythrocytes of land mammals are smaller than in aquatic ones, but their total number is greater. For example, amphibians have up to 0.1 million, reptiles have up to 1.0 million, mammals (horses) have up to 8.5 million in 1 cubic mm of blood. Chimps have an average of 7.3 million red cells per cubic millimetre where humans have only 5.1 million, but chimps have an average of 12.2 haemoglobins per cell but humans average 18.6. (Ibid.) The boost in haemoglobin that transports oxygen to the lungs for aquatic animals is very advantageous for carrying more oxygen through the body. Possibly, the reason for that is aquatic mammals have a reduced number of blood cells, in comparison with territorial animals. Since aquatic animals live in conditions of lack of oxygen when animals are forced to hold their breath, thereby developing the ability to inhale a lot of air in comparison with terrestrial animals. Perhaps the reason people that have an increased number of haemoglobins is an aquatic past.

In addition, there is a lack of evidence from the fossil record to support AAT claims. Most of the evolutionary adaptations described by Morgan would not have fossilized; therefore, many opponents criticize it for no verifiability. Many aquatic mammals demonstrate strong skeletal evidence of adaptation to water whereas humans have a shortage of such adaptations. Moreover, many of the exclusive characteristics of human adaptation can be explained without involving an aquatic phase of evolution by other hypothetical scenarios (Gee 2001). Opponents of the Aquatic Ape Theory note that it creates suggestions without testing them and moreover, it employs purely supporting evidence in the applicable scientific literature whilst at the same time paying no attention to the larger body of conflicting evidence.

Langdon (1997) postulates that movement of hominins from a terrestrial to a marine habitat and back again is less economical than a line of terrestrial hominins throughout the time between 4 and 7 Mya. The hypothesis about a marine habitat has needless difficulties within the sequence of events; this has led anthropology's mainstream to refuse this hypothesis. However, Kuliukas (2011) notes in this connection that the AAT is not a change to an aquatic phase and reverses again, it (AAT) in the main is a stable pattern of hominins living in an aquatic environment more than apes did.

AAT is not a static model; in the last 20 years, akin to any human evolution model, it has developed as a reaction to newly arising data. For example, Verhaegen *et al.* (2002) proposed that all African great apes and Homo developed from a common ancestor that was previously a wading-climbing ape which may inhabited coastal forests in the Mid and/or Late-Miocene, before the Pan-Homo split, in the coasts along the Tethys Ocean that is the modern Mediterranean Sea.

This original concept is contradictory to the Morgan and Hardy vision, who supposed that some ape groups were exposed to a waterside habitat before returning to a fully terrestrial existence. As an alternative to the traditional savannah-dwelling anthropologist approach, Verhaegen *et al.* (2002) argue that a combination of recently revealed Orrorin, Ardipithecus, Kenyanthropus and other hominids' fossil and comparative data now provides evidence showing that most primitive hominids waded and climbed in marshy or coastal forests in Africa-Arabia and fed upon hard-shelled fruits and molluscs. Their Australopithecus ancestors had a similar locomotion but usually preferred a diet including wetland plants, and the Homo ancestors migrated to, or remained near, the Indian Ocean coasts, forfeiting climbing skills and exploiting coastal resources.

Verhaegen *et al.* (2002) believe that traditional explanations for human bipedalism such as aggressive posturing, rising to reach food in trees, looking over tall grass whilst transporting tools or babies are flimsy as no other primates have developed bipedalism for similar reasons. An aqua-boreal locomotion that included both trees and water could have been the characteristic that distinguished apes from monkeys. Verhaegen *et al.* believe that a vertical posture and aptitude to climb with arms raised overhead could help wading primates to leave the water and hold fruits above the water. A well-built body that was easily supported in water and facilitated retention of heat acts to decrease ability for tree dwelling. In addition, tails could not be used appropriately by a wading and swimming primate, they increased water resistance and heat loss and consequently the tail was reduced. They suggest that most hominids might have lived in damp conditions, which were confirmed by new discoveries; they agree that all Australopithecus lived in wet, well-wooded environments and later species existed in more open wetlands. They hypothesise that around 2-3 Mya human ancestors could live close to the coast and use marine products like shellfish and fish which have the long-chain, polyunsaturated lipids that were vital for construction and energizing of large human brains.

Kuliukas (2011a) offers a fresh-water 'River Ape' model of human evolution that is a combination of different models founded upon the traditional anthropogenesis paradigm that unite the Hardy and Morgan ideas with the strong points of those of Ellis, Crawford and Verhaegen. The model underlines that the most important result in the initial evolution of hominins is their bipedality. It states that the last common predecessor of great apes and Homo that has not been found was a coastal or inland swamp forest ape that had optional bipedalism when it was in water but was also a very good climber. It is argued that this is a consequential predecessor to fully terrestrial bipedalism of Homo and to quadrupedal knuckle walking of Homininae. Fossil findings such as *Sahelanthropus tchadensis* and *Orrorin tugenensis* confirm the existence of a common progenitor of Pan and Homo.

The aridification of the African East Rift (4-6 Mya) is thought to have reduced in size the forested habitats, leaving them primarily near water. Later, hominins lived in riparian forests along the wide rivers and lakes of East Africa. It is suggested that during this phase human ancestors progressively moved more bipedally both in shallow water and on dry land. Ancestors of African great apes still lived in dense western tropical forests and assimilated knuckle walking as a post-wading-climbing form of locomotion.

The next phase of the model proposes that, afterwards, some early Homo migrated to coastal regions; conclusively disposed of their arboreal origins and adapted to swimming and diving in the course of the struggle for food. *H. erectus* was successful in developing such adaptations and consequently mastered a range of other niches for its tribe across Africa and Eurasia.

Around 250 Tya ago in East Africa, a vital new transformation in the human genotype caused the speciation of modern *Homo sapiens*. Presumably, it was made by a mutation leading to a gross neural alteration or a hybridization event of two hominin groups which were split around 1-3 Mya in the Danakil depression when the internal sea there dried up. The number of human chromosomes changed from 48 to 46. This efficiently genetically separated the nascent species from its parental groups and resulted in rapid brain growth, full syntactic speech and culture. Following the speciation, probably in Africa, a fully modern *H. sapiens* species was ready to replace all other hominins on the planet (Ibid.).

Conclusion

Supporters of the AAT have proposed two possible ways of anthropogenesis: first, human ancestors had a semi-aquatic habitat, and only later descend to the ground and became savanna dwellers. A second possibility is that human ancestors never lived on the savannas but remained semi-aquatic until the Early Pleistocene. In the last few million years, *Homo* adjusted to a land-dwelling life and therefore many footprints of aquatic adaptation have become moderately hidden. However, some indications still exist and AAT supporters note characteristics that are present in aquatic animals and almost absent in apes and savannah mammals: enlarged pharynx; nose; activity of sebaceous glands; paucity of apocrine glands; direction of hair follicles; a superficial fat-layer bonded to the skin; nakedness over most of the body surface; very large sebaceous glands at the boundaries of the hair and bare skin; thermoregulatory sweating through eccrine skin glands; strong dependence on drinking water but a low drinking capacity; high susceptibility to dehydration in warm, dry and open habitats; rich salty tears; volitional control of breath and a poor sense of smell. The AAT confidently envisages that wading through shallow water was a major factor of the locomotion of the earliest proto-hominids that took to habitats where such wading locomotion was favourable. It also considers that there is a connection between consumption of marine food and development of the human brain in all periods of human evolution.

Despite the fact that there are many arguments supporting the aquatic ape theory, they are not adequately influential to subdue the opposing arguments. Many critics of the AAT note that in comparison with the Savannah Theory, it does not permit the careful clarification of some inconvenient facts and it cannot give better predictions of new findings than the traditional approach. Proponents of the AAT do not develop hypothetical explanations that are tested by the traditional approaches of natural sciences and that is the weakest point of the new theory. The AAT is unclear in both absolute and relative dating. It emphasises various subjectively chosen mysterious characteristics of Homo-sapiens and offers purely theoretical explanations for them without any attempt to provide comprehensive, causal explanations for the origins of these qualities.

The concept has a deficit of precision and, as a result, does not have an adequate amount of testable hypotheses and associated prognoses. Moreover, it does not give more or less complete consideration of the information about human evolution than researchers already have. The arguments of the AAT represent an unverifiable narrative of explanation of the evolutionary development and the theory is far from the modern discussion in evolutionary biology regarding the importance of apparent adaptations. However, the aquatic theory continues to be an open question and as a minimum must give motivation for further research as a 'good detective follows up the least promising clues as well as those which seem to point to a simple solution' (Westenhöfer, 1942, p. 150).

Chapter 4

Anatomical and physiological

exclusives of humans

Human nature

As the founder of philosophical anthropology, that is a phenomenological investigation of the human being and his place in nature, Max Scheler distinguished two opposing concepts of man: man as a biological creature with a fixed place in the natural animal kingdom and man as an existence with an apparently abnormal place in the world with a specific essence. In his last paper Man's Place (1928) Scheler asked the questions 'What is man?' and 'What is man's place in the nature of things?' His modus operandi is discordant with the simian theory of human evolution that has dominated since the 19th century and postulated in the canonical form that a man and four-armed monkeys originated from a common ancestor.

He noted that the concept of man is alluded to in ambiguity, on the one hand, it signifies the particular morphological characteristics of man as a mammal, but on the other hand it signifies a set of characteristics that must be distinguished from the concept of animals including other mammals. Man is an exceptional combination of vital and spiritual being, both part of and yet different from animals. Scheler divided all psychophysical life into few stages of rising complexity: vital being, instinct, associative memory and practical intelligence. The highest stage, practical intelligence, is the ability to respond meaningfully to a new situation without trial and error and Scheler believed that such a capability of intelligence is characteristic not only of human beings but also of apes. He supposed that there is no vital qualitative distinction between humans and chimpanzees; on the psychophysical level, there is nothing unique to humans. The new element that gives man unique characteristics and essential nature is a phenomenon that cannot originate from the natural evolution of life. That phenomenon, spirit, goes beyond psychophysical life is an element not shared by animals, and that gives man the ability to behave autonomously from his desires. Spirit is independent of man's physical organism; it cannot be studied by biology or psychology (Ibid.).

Scheler solved the identification puzzle of human nature theologically, contrasting the essence of man as the image and likeness of God, with his unreasonable animal nature, which only had a chance to be realised thanks to divine support. However, this opposition of *Homo Naturalis* to historical man involved in the divine principle contradicts traditional evolutionary schooling. In contrast, the idea that the social is not only a denial of the biological, but also its consequence, was actualised in the works of ethologists Lorenz and Tinbergen. They managed to trace the commonality of behaviour between animals and man, but these studies of behaviour did not lead to a solution of the problem of human nature and its origin.

Numerous attempts to identify man and his society with other species such as how human society is similar to the lion pride, or a pack of wolves, or an ant family, or to diffuse chimpanzee communities and so on, contain a logical error in substituting the base of the comparison, when one species is compared with many others simultaneously. Consequently, there are similarities everywhere, but no complete analogy with any other species. In this case, there is comparison not of holistic human behaviour, but of insignificant parts thereof with the social behaviour of other species.

This produces a negative result in such attempts to equate people and animals. However, it leads us to the idea of the special nature of man, whose manifestations of social qualities are virtually infinite, in contrast to other species where everything is firmly regulated by instinct. Lions will never be social like a chimpanzee, and bee society will not be like that of lions: the drone will not become the king of the hive and the lioness will not be the queen of her pride. People can have diverse societies such as the lion pride (for example the Muslim harem), or the diffuse community of the monkey (polyamory), or like swans (monogamous pair bonds), or the communistic ant colony. Both man and woman can be a human leader. In addition, a man can be isolated, entering into only sporadic contact with the aim of satisfying his animal needs, leading him to the limitlessness of asociality, outside the cluster of social animals. Again, in this cluster there are analogies in the sporadic contact of tigers, bears and crocodiles.

Sporadic contacts of animals for various species are diverse. In some species, the initiative is manifested by a male and in others by a female;

but people of both genders take the initiative to various extents dictated by culture and personality. Moreover, a person can be a hermit who avoids sexual contacts, even those associated with reproduction, which in animals is impossible in principle. This leads the person to a position where the animal property is absent except at the anatomical level and all behavioural comparisons become meaningless.

Therefore, it can be said that the achievements of ethology did not lead to the solution of the problem of filling the gap between animals and man. They could not answer the question posed by Scheler of how a creature so poorly physically adapted to the wild environment, almost naked, could survive whilst leaving a comparatively safe tier of the tropical forest and walking on the ground on short, crooked and weak legs.

The search for an adequate methodology

If the truly enormous factology of human evolution is not built into something intelligible, the issue is of a methodological nature. Therefore, it is useful to take a step back in order to analyse the methodology. The misconception was at the very beginning when the monkey was subjectively assigned as man's predecessor based on external similarity. Therefore, the concept of the ancestor should be the result of a scientific investigation, not its beginning. Scheler, having tried to find a solution, gave a theological explanation rather than a scientifically monistic solution. Therefore, the search for a methodology capable of proposing a lead to the ancestor of man turns to influential classic researchers.

In the 1830s, Scottish geologist Charles Lyell developed the concept that all features of the Earth's surface are generated by physical, chemical, and biological processes through extensive periods of geological time. He formulated the Uniformity Principle, which declares that past geological events must be explained by causes identical to those now in operation, so the present is the key to the past. The methodology of Lyell gives an idea for understanding the origin of man. If you want to understand how a process occurred, but do not know where it started or how it developed then you can take the end of the thread, without imagining where its beginning lies. The end leads to the beginning, and not vice versa. It is simple, like unwinding Miss Marple's ball of wool. It is not by chance that the perceptiveness of the amateur consulting detective of Agatha Christie's crime novels is associated with the image of the unwinding ball. However, paleoanthropologists of the 19th century did not start from the end, but from what they considered as the beginning. As a result, their efforts have tightened the chaotic anthropological tangle, which they cannot unravel. They started with the monkey, whereas it was necessary to start with man.

The principle of Lyell means that in evolutionary research it is necessary to isolate the most developed form and go backwards into the depths of time. Other ways of research give more questions than answers, often unscientific and turning into empty fantasies. Science is unwinding the coil, so often the incorrect scientist tries to push the hypothetical principle he invented himself. The tangle of threads is much easier to unwind from the end.

Karl Marx named this way 'the ascent from the abstract to the concrete.' He supposed that man descended from the monkey and he thought that human anatomy is the key to understanding the anatomy of a monkey. It can be noted that this is the path from the end to the beginning, and not vice versa. His theory of socio-economic formations, he based on political economy that is a doctrine of how commodity circulation changes social relations. At the same time, Marx began with the analysis of the formation, where commodity circulation was highly developed, from capitalism, distinguished the main specific features, which became the initial principles of the theory. Having created the theory of the development of socio-economic formation, he restored the whole sequence of formation from capitalism to the hunter-gatherer societies based on egalitarian social relations and common ownership (primitive communism). After creation, the theory was backfilled with content, developed to a level of detail. It all began with a selection of abstractions based on the most developed, actual form. Thus, one can see the progression of the analysis from the final point to the initial.

Charles Darwin employed a similar method in the book 'The Origin of Species', where he primarily considered the artificial selection applied to man, which is the latest and most developed form of selection, then he pulled out the principles of selection and, on this basis, he reconstructed the process of natural selection that nature conducts. Consequently, he created a general theory of the origin of species.

Actualism and anthropogenesis

The traditional theory of anthropogenesis does not begin with the justification of the methodology of scientific research, but with eclectic descriptions of the bone remains of hominids in Africa. In this case, each author builds his own anthropoid evolutionary trees, the multiplicity of which is vivid evidence of the eclectic approach. In fact, traditional paleoanthropologists, going from ape to man, and not vice versa, created a kind of ideal, constructed primordial creature that is half-human and half ape, called it the missing link and searched for it all over the world. In appearance, it looks like the science of the past, but in essence, it is pure futurology with a strong flavour of science fiction.

Evolutionary scientists seek characteristics of current humans based on those of the past, with the purely speculative construction that the answer to the future is in the human past. Texts of modern paleoanthropologists have become almost esoteric tomes, the territory of streamlined phrases and oracle-like definitions. Many people intuitively feel that there is some sort of error in this theory, and as a result their concept appears in isolation in today's society. Therefore, one can use a different approach: it is necessary to distinguish in the human body as many as possible of the features that differentiate man from animals, and in this way investigate where and what were the origins of human nature. These simple features, that Ten (2005) names the anatomical and physiological exclusives (APE), should lead researchers to the landscape in which man was formed and to common ancestral species, and the factors of evolution which worked upon the creatures from which humans evolved. In this regard, it is reasonable to use the above principle of actualism justified by Lyell, since the natural history of Earth is the same evolutionary concept as the theory of anthropoid evolution: past evolutionary events must be explained by causes identical to those now in operation, so the present is the key to the past.

There is an obvious difference between religion and science in the methods of cognition. Religion invariably starts of from the beginning, which is a pure declaration, because the beginning is hidden. 'In the beginning was the word', says the Bible (John 1:1), and then follows a sequence of unverifiable statements. One can believe, but no one can verify them. Whereas science begins with the end, namely with the obvious, with the empirical present and goes deep into the theoretical and the past, unwinding the tangle. According to this criterion, the simian theory is rather religion-like as it began not from the end (from modern man), but from the unsubstantiated statement that is considered as the beginning. As a result, the theory got confused in its contradictions, as many religions are confused. Its apologists defend the four-armed ancestor, without serious arguments, as theologians defend religious dogma.

Ten (2011) offered to withdraw from psychologism in the construction of the theory of anthropogenesis, to take as the point of reference the corporeality of modern man, and analyse it from an evolutionary point of view without prejudging conclusions based on external similarity. In evolutionary biology, factors of similarity sometimes do not work and generate errors. Seals are like tuna in many respects, but these are very different classes of animals. In the Amazon, there is a capybara that was previously considered to be a deer because of its large size but is actually a rodent. In Africa, there is a small burrowing animal Hyraxes, which was considered a relative of the ground squirrel but later was considered a close relative of the elephant.

The scientific theory must begin with a methodology, not directly with facts alone. Facts cannot create a theory as they can be eclectically manipulated. A logical, consistent system of evidence can be built only on the basis of a unified methodology and not by overreliance on individual observations.

Ten (2011) adapts the principle of actualism to the theory of anthropoid evolution by systematically attempting to identify in the human body such anatomical and physiological exclusives that distinguish it from monkeys and other animals and to denote the biological specificity of man. The aim was to determine the type of animal man belongs to: arboreal, marine or some other type. For this, it is necessary to reconstruct the landscape in which the human body formed, to get an idea of the conditional factors of anthropogenesis. Various authors since the beginning of the 20th century, for example Hardy, have drawn attention to the fact that man is a naked animal and has subcutaneous fat like water mammals, so they suggested that man be considered a descendant of the water monkey. However, if we are talking about adaptation to an aquatic environment the question arises about why the monkey is immediately attractive as a predecessor, and why these exclusive adaptations cannot be considered objectively.

Thus, the process of creation of a new evolutionary theory has to include a definition of a limiting evolutionary form of an object, creation of the preliminary scientific model based on the studying of the limiting evolutionary form. Then the process of evolution is reconstructed and intermediate stages based on the logic and historical analyses are defined. The emergence of the most developed last form receives concrete justification; that is, the end coincides with the beginning. The anatomo-physiological exclusives allocated to various authors will be analysed further so that in the following chapters it is possible to start creating a model of human evolution in order to combine the theory with the landscapes where these species could inhabit. Having determined a suitable anatomy and landscape, one will be able to identify possible ancestral species to try to find answers to the eternal questions of humankind such as how the consciousness of the human race arose.

One of the first to identify the individual anatomical and physiological exclusives of man and study their somatic base was the German anthropologist Arnold Gehlen. He noted that a man does not have mammals' wool or fur that is essential even in a warm climate and consequently there is no natural protection against temperature changes. His sensory faculties are not developed in comparison with animals; the period of childhood is relatively long, needing very long-term care in comparison with other animals. However, since man is viable, the preconditions for the solution to this problem must be in man himself. The possibility of realization of this hard-earned achievement of existence must come from the general structure of man (Gehlen, 1988).

He noted that all the specific human abilities should be related to the solution of the question: how can such a non-adaptive being as man, such a weak creature in comparison with other animals, be viable? To answer this problem, Gehlen proposed to start from the universal structure of man and not compare the physical nature of man with the physical nature of chimpanzees.

Simple questions and difficult answers

Beginning the analysis of human anatomical and physiological exclusives we will try to find answers to simple questions because the more simply the question is posed; the more chance it has to be solved. Let us think of elementary questions, for example, why does a human have a unique type of hair growth, which grows continuously and can reach an incredible length? This is an absolute anatomical and physiological exclusive of man.

Man is the only animal whose body is almost naked and yet his hair grows continuously. The Guinness Book of Records recorded a length of 28 metres. There are no such monkeys and could not be, a monkey would not have survived with such hair in the forest or the savannah. After all, short hair on the head is enough for protection from the sun, like monkeys and other tropical animals have. There is no good explanation of this strange exclusive and its evolutionary significance. The second of man's absolute exclusives is sweat glands located all over the body, including the feet and hands. The need of the sweat glands in these places is completely incomprehensible in terms of the traditional theory.

The third APE is the incredible intensity of the sweat glands. Most animals, such as birds, do not sweat at all. Others, such as monkeys, sweat very little. Others sometimes release a lot of sweat due to excessive physical stress, for example, the exhausted horse, which then often dies. Only man sweats continuously, even in comfort and complete inaction. In such circumstances, the body releases an average of 0.7 litres of sweat per day. In hot weather, a person produces up to ten litres of sweat. Now imagine our ancestors in the wild jungle or savannah. Almost all predators search for victims by smell. This sweating would make our ancestors easy prey that could not hide anywhere.

Why do we sweat so much? The simplest explanation is thermoregulation, but it is ineffective thermoregulation for life in the forest or savannah. Sweat does not protect against cold in any way. At night, even in the tropics, it is cold and animal fur is not a superfluous adaptation. Sweat also protects from heat very poorly. Fur is a better protection for all animals from both the cold and the heat. We suffer from the heat when it is over 25 degrees Celsius. Being naked, humans freeze to death when the temperature is only 0 degrees whereas furry animals feel comfortable. The temperature range that an almost sweatless camel can easily survive is around 100 degrees from minus 50 to plus 50. For a naked man, the comfortable range is much smaller: from plus 18 to plus 23. Outside this range, man feels either cold or hot. This suggests that people are not well suited to the heat but to a moderate, stable environment of temperate conditions. The comfortable range for humans coincides with the temperature range of the subtropical sea with its stable conditions.

Cold sweat is the fourth absolute human exclusive. The warm sweat may be explained by the thermoregulation of body whereas cold sweat, which instantly appears in a moment of fear, covers often even the feet and hands and does not have a good explanation. By the way, unlike humans, monkeys do not release such sweat.

It is interesting to understand the nature of human ancestors with nostrils directed downwards. Having lived in the forest, it is necessary to constantly sniff the air to smell a predator, therefore animals' noses point forward and slightly upward. Human noses pointing down are nonsensical if we construct a theory of human origins from forest monkeys. Forest monkeys' nostrils point forward, giving the possibility to sniff the air with each breath without lifting up the head. The nostrils of man are another exclusive, constituted in the likeness of a deepening bell. Especially clearly, the bell shape is detected while diving when the person instinctively straining nose, extended in the external part and narrowed in the internal, the nostrils create an air lock. Water cannot overcome it, so people can dive, withstanding pressures up to four atmospheres (20-meter depth). There is no other land animal capable of this. Also, the poorly developed ability to smell brings humans closer to marine animals because terrestrial animals, as a rule, developed a sense of smell much stronger than that of sea animals.

Absolute human exclusives

Lindblad (1987) proposed a theory of the origin of man, which modified the concept of Hardy and which assumed that man evolved from some forest monkeys that had a semi-aquatic way of life in fresh water. In addition, he selected 15 human anatomical and physiological exclusives, some of which were already offered by Morgan (1982) and he constructed the systematization of APE. However, Lindblad used a narrow base in this systematization: distinguishing between human species and pongidae that is an archaic primate taxon including the gorillas, chimpanzees and orangutans.

The list of APE of Morgan and Lindblad has some shortcomings. The problem is that many of the features are deprived of objectivity. For example, some APE such as a vertical posture, large skull, large brain volume, small canine teeth, flat face, high palatal vault, agile tongue and forward protruding nose are marked by a more or less subjective principle. The allocation of human exclusives is made under high demands: an exclusive either is or is not. An example of a mottled, less maximalist, approach would be describing pre-Homo human ancestors as having somewhat vertical posture but in this case, any branchiate that moves sometimes vertically and sometimes relying on the hands will possess this human exclusive. In addition, in this context, the size of the skull cannot be a beneficial example of a human exclusive due to its large variability.

In such APE as small teeth, especially canines, the quantitative criterion is not essential, when qualitative description is possible, for example associated with the lack of human diastema. Some exclusives can contradict with others, for example flat face contradicts with a protruding nose where both are exclusives. With regard to the human nose, it has an absolute human exclusive - the nostrils directed downward. The tongue, noted as APE by Lindblad, cannot be accepted as an exclusive as apes have tongues, which move as in Homo, but in this case it is more productive to analyse a human nasopharyngeal system as an absolute exclusive. In this situation, with regard to the monkeys, one can talk specifically about mouths associated with the forward extension of the entire lower part of the muzzle and the superciliary arcs.

The APE include short arms, long legs, foot not opposed to the big toe, palm with a thumb opposable to the other four fingers, reduced hair growth on the body (implicitly suggesting reduced hair in man, although a model where hair appears as a result of adaptation to land-existence can be built), sweat glands throughout the body, a new type of adipose tissue (especially in children), a large penis in men and large breasts in women, long hair on the head and underarm hair (it is necessary to speak not about the length of hair but about the type of scalp hair since this hair normally grows continuously, unlike armpit and pubic hair, and can reach an incredible length).

In addition, Lindblad's basis of classification of APE is not clear. Different researchers highlight features that differentiate humans from our nearest animal relatives, namely pongids: relative exclusives that distinguish humans from different aquatic and terrestrial animals and absolute exclusives that are anatomical and physiological peculiarities, which no other species has.

Lindblad and supporters of the aquatic theory explain opposability of the thumb by primitive people washing and eating protein-rich shellfish by hand, drawing an analogy with the raccoon. However, raccoon limbs have long claws, with which they harrow silt and sand, searching for hidden crayfish. However, the hands of man end in nails, less durable than these claws and the likelihood of breakage of these nails while hunting is very high. One can see again ambiguity in the evolutionary interpretation of exclusives.

Sex of primates and humans

The size of the penis has also become a subject of debate among the supporters of the aquatic theory. A male gorilla, despite having a much larger body size in comparison with man, has a 5-centimetre phallus. In humans, it is longer because it is said, by Morgan (1982), evolution moved the vagina deep into the body, probably for better protection against salt water and sand scratching. As a result, the penis was forced to increase in length. This is for almost the same reason that the neck of a giraffe stretched. As the giraffe lived in a very arid area, where almost nothing grows on the ground, they were forced to look for food in the trees and make constant efforts to reach it. 'From this habit long maintained in races, it has resulted that the animal's forelegs have become longer than its hind legs, and that its neck is lengthened.' (Lamarck, 1809, p. 256). This explanation can be accepted as part of Morgan's implied growth of the penis in the process of evolution from ape to man, and not its reduction in the reverse process from man to ape. Morgan and her supporters do not even consider reduction of the penis, despite the fact that there are many purely terrestrial animals with giant phalluses. Nature always manages to reduce what is superfluous more easily than to create something new. Generally, the size of the penis is a very narrow exclusive separating people only from anthropoid apes. From land animals, as such, it does not separate us so the aquatic factor does not work here. The phallus of horse, dog, and boar are quite impressive in relative size and, in fact, these animals prefer to mate on land.

The mating patterns of apes are different from man. Orangutans live in trees; males do not fight over females and do not care about them, nor about the cubs who after four years leave in separate semi-mature groups. Gorillas live in groups in the forest; one of the males is completely dominant but has no jealousy and allows subordinate males to mate with the females. Males do not perform courtship displays to females, and neither do they feed cubs and they drive small cubs away from themselves.

Chimpanzees live in a more open landscape and spend more time on the ground. Their groups are more extensive and their relationships are warmer and more varied. The males do not form a strict hierarchy, are not jealous of females, do not do mating calls and do not feed their offspring. Mothers, sisters and older daughters help to care for little cubs. External genital organs of both male and female chimpanzees have larger dimensions. The erect penis of the male chimpanzees is larger than those of other apes, but less than of man. However, the share of the weight of the testes in relation to the weight of an individual chimpanzee is quite high, averaging 0.269% of body weight, which is the highest figure of the primates. The relative weight of testes in gorillas and humans are 0.017% and 0.079% respectively (Short, 1979).

The large size of the testes reflects their high spermatogenetic capacity that according to Short (1979) is an adaptation to high frequency mating within the community of chimpanzees, in which the number of females is usually more than that of males, and mating occurs throughout the whole year. One can also note the short duration of intercourse for such large animals and a small number of frictions, an average of seven (Ibid.) required for ejaculation. Practically, a male in a few seconds and a few frictions performs sexual intercourse. Often they show no emotions before and after intercourse and they do not have any love courtship. One can say that all female apes mate rarely and irregularly, they are hypo-sexual and quite powerless whereas males are not jealous and they do not take care of their females and children. In humans, there is a strong dependence on sex. It can be assumed that the sexual activity of our ancestors occupied much of their free time. Despite the difference in the sexual behaviour of apes and man, both have menstrual cycles of females that replaced the annual ovulation of other primates.

Many specialists consider that the original family structure of human ancestors in times of forest-dwelling lifestyle consisted of a male, one or two females and children. In feeding areas, families combined into groups. Grown-up children of both sexes were expelled. The existence of monogamy is evidenced by the presence of the instinct of jealousy that is absent in monkeys with the group forms of sexual relations.

A man mates all year round while the vast majority of terrestrial animals have a seasonal cycle of mating. However, a few other mammal species copulate out of season, for example, domestic animals and human-satellite animals such as the rat and mouse. In this case, it is necessary to talk about the effect of domestication. In wild conditions, muskrats and field mice breed by a seasonal pattern. People are the exception in the world of animals on such grounds as sexual behaviour, since the desire for sex and its frequency is far greater than the biological optimum; sex became a self-contained sphere of life, separate from reproduction. In animals, a pregnant woman, starting from the moment of successful impregnation, does not feel the craving for sexual intercourse with males. In addition, after childbirth asexuality is always present, sometimes for a very long time, until the children grow up. All animal females, other than humans, are able to mate only for a limited period of time or only at a time of ovulation, or some time thereafter. It turns out that animal sex is strongly associated with fertility. In contrast to animals, a woman has a unique ability of continuous sexual intercourse, even during pregnancy and during breastfeeding when a woman's body is in a quite different hormonal situation. The absence of external signs of ovulation is also unusual. Only dolphins have similar behaviour; they are capable of year-round sex and they have approximately the same cyclic recurrence as women. Female dolphins are able to mate all year round; they do not have a long oestrus of the type of terrestrial animals.

The theoretical concepts that explain hypersexuality are relatively few, primarily because researchers do not agree with the idea that hypersexuality itself is a human exclusive. However, Darwin in his book 'The Descent of Man and sexual selection' attached great importance to the action of sexual selection. Man, in the course of evolution, has gone through a period of intensive sexual selection, which possibly formed hypersexuality. Enhanced sexuality is observed in some animals living with group sexual relations such as in the vervet monkey. Female vervets, as typical non-human primates, ovulate in sync with each other once a year; nevertheless, there is no external indication of this. Moreover, vervets are able to mate from two months before ovulation until the second half of pregnancy. Over such a long period, a female has time to mate with 60 per cent of the males in the group and all males share food with her. One can say that due to female hypersexuality with a greater number of males, females can get food for themselves and their cubs. Not knowing the timing of ovulation, all males consider her babies as their own and if one of the fathers dies or moves to another group then the babies do not remain fatherless. Natural selection, using hypersexuality and the hidden ovulation of vervets thus managed to overcome the typical full domination of males, extending the time for partial female dominance and thus providing reliable care of the female and her children. This arose from the fact that they have to dwell in the open landscape, where the female has difficulty in producing meals in the required quantity.

Ten (2011) considers that high human sexuality is associated with features of the nutrition of human ancestors, who consumed large amounts of molluscs, having a high frequency of mating and the possible existence of specific genes of sexuality. Although this hypothesis looks heuristic has not found confirmation, it is an interesting explanation of the sexual energy of the great apes being not so high in comparison with man.

More interesting still as an explanation of relatively low sexual energy of the great apes is Ten's belief that if chimps and gorillas had big penises and prolonged friction of sexual intercourse, it would require a greater excitation for arterial blood flow in the cavernous body of the penis. That in turn would require sex games, more eroticism and sexual love, which is associated with preferences, with desires, with the competition and ultimately fights for females. All of this could change radically the life of simian flocks. By nature, chimpanzees sufficiently conflict with extreme violence in fights. If there were also fights for females it could be a problem for the survival of the species. Evolution simply swept away this cause of conflicts. One can say that a small phallus, in comparison with both sea and land animals of the same size, of the pongids is associated with their specific sexuality.

Humans have hyper-sexuality and this factor played a major role in their evolution. Human sexuality, in the process of anthropogenesis, has also changed as the survival of the species in harsh environmental conditions required elimination of conflicts of hypersexuality so they became regulated by social rules. People were not able to return to typical seasonal sexuality of animals due to the irreversibility of evolutionary change. According to Dollo's law of irreversibility (1893), evolution always moves forward and an organism cannot go back to the original view, from which it originated, even though it returns to the original conditions of dwelling and once lost in the process of evolution organ an cannot be recovered during subsequent phylogenetic development.

In this context, one can speak about the phenomenon of involution, that is the reduction in the evolutionary process of individual organs such as the ape's penis, to simplify its function. In other words, this is the opposite of organ development. Engels in his 'Dialectics of Nature' one of the first philosophers that admitted the possibility of involution, speaking about the lowest savages who could return to a more beastlike existence. So here, one can say that the development of the great apes at a certain stage of development occurred involutionally from the more developed human precursor to the less developed species, forming modern apes. Such an approach can easily solve the problem of the genesis of the chimpanzee.

Human hair mystery

In the previous chapter the aquatic theory was presented but there are many questions about the interpretation of the various exclusives made by supporters of this theory. For example, Hardy (1960) believed that long hair played a big role in the care of offspring. Children swimming in the water were able to cling to the long hair of parents in order not to drown. It is probable that hair could play such a role, but children born in the water often swim better than adults do. On the other hand, it is difficult to explain the dedicated effort of the human body to the preservation of the hair, distinctive to Homo. Only factors that have a direct bearing on the survival of the organism should cause such selflessness as when, during famine, a brain sacrifices internal organs for the sake of maintaining external hair cover. Ten (2011) supposes that this evolutionary phenomenon can be explained by the fact that at the beginning of anthropogenesis a man deprived of hair died, and moreover if a woman did not have hair her children also died. Therefore, hairless women are rare in modern times, as they left no offspring. Modern men often lose their hair but usually not until the age of twenty years or later, which in ancient times was almost the average lifespan.

Nature is primarily concerned with security. It gives animals the ability to change colour (chameleon), to apply a smell (skunk, ferret) or a masking liquid (octopus), to jump out of the water (fish) or to dive into it (birds). Man was deprived of effective means of self-defence and running speed, but he was granted salvation by hair that grows continuously at an average speed of 13 centimetres per year. Let us imagine what was most dangerous to our ancestors. It was not the sea predators such as tiger sharks, from which people could escape in the land and was not the land predators from which people can escape in the sea. Therefore, perhaps the greatest danger to the early hominids could have been snakes and predatory birds. In ancient times, there were flying predators capable of lifting a man into the air. For example, the Maori legend describes the monstrous bird pouakai that was an extinct species of eagle that once lived in New Zealand in the Middle Ages. With the wingspan of more than three metres, the eagle was not a scavenger, but a predator with claws not inferior to those of the tiger.

From such a bird, it was not easy to escape in shallow seawater due to the high speed of its attacks that reached 300 kilometres per hour. Adults or children could not escape but they could submerge in the water, scattering the hair forward with one hand movement so it is fluttering like seaweed, disorienting a predator and hiding its owner. A camouflage device like human hair is observed only among marine creatures. The most famous of them being the squid and jellyfish. Long thin tentacles, waving like seaweed, are used as masking and to capture prey.

Scalp hair, unlike fur growing on the body, has a temperature and smell not of the body but of the environment. Therefore, they also helped to escape from major type of snakes such as anaconda that are dominant where human ancestors originally lived, oriented not only by the infrared radiation, but also by relatively undeveloped smell. Seeing the snake, the man sweated a cold sweat, thereby lowering the body temperature, squatted and with a sharp movement of the hand threw the hair forward, creating a temperature/aromatic veil. For a short time, this significantly reduced the visibility of the man to the snake.

The protective camouflage function of hair could be performed due to periodic constant contact with warm water. Hair with greasy secretions of sweat glands, having a specific pungent smell, would have unmasked and betrayed its owner. What was an evolutionary advantage in a water-earth lifestyle became a defect in hot rainforest and sultry savannas. Hair became in the forest not only unnecessary but also dangerous for life. Human ancestors could not survive in the forest and the savannah with this flaw, especially if we take into account the fire risk of long hair. Thus, living in the forest people had badly smelling unmasking hair that was an obstacle when running in the forest and the savannah; it was a fire hazard and could host various pathogens.

Each of these negative factors of hair could alone destroy an entire population of human ancestors. Certainly, our ancestors could get rid of the hair using fire, a long period of grinding, of by pulling it out. However, to do that they had to first gain intelligence and wield fire. The gorilla has a humanoid gesture; under stress it scratches his head. However, this gesture is not her invention but a reflex, inherited from hominid ancestors. In stressful situations, people's hands also involuntarily reach for the back of the head, because our ancestors over millions of years of evolution, many times a day had to throw by hand a bush of long hair forward.

Tan, fat and tears

Another human exclusive is crying with abundant eye lacrimation. No other animals are able to cry. Physiologists explain this phenomenon with reference to children who attract their mother's attention by crying. On the other hand, it is possible to attract attention by squeaking without tears as animals do. Although one can accept this explanation, it does not explain the phenomenon. In addition, it does not explain the tears of adults who can cry in silent sobbing. Tallis (2010), by way of an explanation of adult crying, supposes that adults keep their children's anatomical and physiological characteristics until old age. That is, one can talk about the existence in the evolution of man of paedomorphosis (juvenification). This is evolutionary change in organisms characterized by a complete loss of the adult stage and a corresponding shortening of ontogenesis, in which the adults of a species retain traits previously seen only in juveniles (larval stage). More detail about paedomorphosis will be discussed in the next chapter.

As we noted early, in terrestrial mammals the fat deposition initially occurs in the abdominal area and subcutaneous fat in normal circumstances is not typical for animals of this type. Terrestrial mammals, dressed in fur, have no need of a layer of subcutaneous fat. On the contrary, in marine mammals, even those who lack food, subcutaneous fat is necessary for thermoregulation. In humans, one can observe the accumulation of adipose tissue by the marine type: firstly under the skin and then in the abdomen. Even with a deficit of food, the human organism lays some fat under the skin. Of course, in special circumstances, such as in a zoo, one can meet animals, with much subcutaneous fat. In this case, it is an energy reserve, which the body produces from abundance of nutrition. There are land animals that rapidly accumulate subcutaneous fat, for example pigs that, like people, formed in a mixed aqueous-terrestrial landscape. This again confirms the idea that different species become similar due to similar conditions of formation.

The pigment melanin is produced in a specific group of cells called melanocytes. It is produced by the oxidation of tyrosine followed by polymerization. In the process of melanogenesis, melanin is distributed in the skin and hair. White people have low basal levels of melanogenesis but UV-B radiation causes an amplified melanogenesis that protect the hypodermis from this radiation. The dark colour of melanin absorbs a greater part of the UV-B radiation and so prevents the penetration of the UV rays into the body, as large doses are dangerous. Therefore, the more ultraviolet radiation hits the skin, the more intensively melanin is produced. The ability to tan is one more absolute exclusive of human anatomy and physiology. There is no tanning in any other animal. For example, almost naked white pigs cannot change skin colour. Their skin may briefly turn red from the sun's rays, but will quickly become white again. Many animals can produce melanin as a pigment in other cells called melanosomes, however, but they do not have ability to block UV-B radiation.

Humans have a high demand for iodine of a similar level to the marine mammals and therefore people suffer from diseases specifically related to iodine deficiency. No other animal on the planet has similar illnesses and this shows a connection with residing in the marine environment in the past. Marine animals, which also have a high demand for iodine, do not suffer from iodine deficiency because it is readily available in seafood. Land animals do not suffer from iodine deficiency because it is not necessary to them in such large quantities as marine mammals and people.

Human ears, lips, tongue and chin

The form of the human outer ear is similar to a bell and creates an airlock when immersed in water, also a human anatomical exclusive. Such an adaptive ear shape allows the assumption that its evolution occurred in an aquatic environment. The majority of land animals, including monkeys, do not have these adaptations; they have an ear with a meandering cavity but no hemisphere. Marine animals have more developed and reliably isolated ears for diving and this allows them to dive deeper than man with his transitional type of aquatic adaptations, an exclusive mechanism for living at the boundary of land and water.

A human has a hole which connects the nasal and oral cavities. The human larynx and oesophagus are, in their beginning, one tube whereas most animals, except species from Cetacea, do not have a fistula between the mouth and blowhole; they have two different pipes. This is crucial for human survival. When a man dives, some water can get through the nose but it falls through the oesophagus into the stomach and it does not get into the lungs. Human nasopharynx is a unique and exclusive tool.

Many animals have lips but there are differences between the lips of ruminant animals (such as horses and cows) and those of humans. Ruminant lips are designed to pluck grass and are made up of exogenous tissue, denser than muscle tissue. Human lips are endogenous, soft tissue, which are simply turned out mucosa of the mouth and therefore need constant wetting. Such lips are another absolute exclusive of human, it is important to note that the apes have no lips; they lost them during evolution or maybe did not acquire them.

The very agile tongue and bulging soft lips of humans, different from those of animals, could have been formed during the foraging of shellfish in coastal shallow water. The human tongue is perfectly adapted to lick the contents of a small shell and the lips adapted to suck these contents. Later in the process of humanization, they were used for speech. Without lips, it is difficult to suck an oyster or flesh from the claws of a crab. Tissues surrounding the mouth of most mammals are typical of the skin of the animal and have less elasticity than inner tissues that nature wrenched out in humans in order to create a biological pump. Lips act as an elastic gasket. Evolution, by creating lips, made possible speech because a man speaks not only through the tongue but also because of the mobility of the lips that articulate speech. Eating shellfish was possibly one of the first forms of the labour of man as he had to work not only with his hands but with his brains due to the enormous variety of molluscs and for each type he had to find a unique approach (Ten 2011).

Humans have a pronounced chin projection. There are no species of monkeys, including fossils, that have anything like that. Modern anthropologists have difficulty explaining how and why the flat face of a monkey might acquire chin projection. However, it may be recalled that during swimming or standing in water up to his neck a person pulls the bottom of the face forward and upward. Perhaps this chin projection was formed over millions of years of such physical exercises and it prevented the ingress of water to the mouth and airway (Ten 2011). The chin development was connected with the survival of the species. People who did not have enough chin projection quickly choked in water and such knowledge of our ancestors is very firmly cemented in the unconscious. As a result, though having no obvious sexual or gender-based importance, the chin plays an important role in the attractiveness of men and women in almost all cultures.

All terrestrial animals, when uttering sounds, use an inflow of air because with the exhaled airstream into the atmosphere from the larynx would be carried odours, including harsh aromas of decaying meals. Generally, animals avoid leaving extra traces of their presence, unless it is associated with self-defence, designation of the territory, or sexual behaviour.

Animals of the sea physiologically cannot use an inhaled airstream, because of the risk of death of accidentally breathing water into the lungs. Changing the airflow direction during sound production is an important adaptation and additionally validates the human marine origin. Later, this method of emitting a sound with an exhaled jet of air played a huge role in anthropogenesis because articulate speech is possible only using an exhaled air stream. This circumstance is serious evidence that evolution of man could not take place from monkeys as Hardy thought because speech and thinking are one indivisible phenomenon. Despite the simplicity of the proposed hypothesis, it provides a logical and harmonious explanation in comparison with the traditional theory of anthropogenesis with its endless strained interpretations and incorrect assumptions.

The human foot

As we noted earlier, many animals are able to stand briefly on their hind legs and all of them have the primitive flat foot. Only man has a complex spring mechanism of the foot. Many anthropologists note that it was formed due to the transition to bipedal apes. However, it is harder to imagine a flat foot evolving into an arched one than it is to imagine the inverse process. Physicians insist that platypodia is not corrected by walking; moreover, it can be aggravated. Ten (2005) believes that the only intelligible explanation for the appearance of the springy foot is swimming. In the water, Homo strained the feet so that they become similar to semi-circular fins. Over millions of years, this physiological stress appeared entrenched in such an anatomical exclusive as the arch of the foot.

All animals have a skeleton that is necessary and sufficient for success in life without comorbidities and exceptionally man has different orthopaedic diseases of the various joints due to the 'undeveloped' nature of the skeleton for land lifestyle. It can be explained that in terrestrial mammals, the mass of the body, relative to the skeleton mass, is 2-3 times greater than in sea animals. For example, the goat, which is comparable in size to man, has a specific weight of the skeleton without horns of 16%; the dolphin has only 5%. Man has a ratio between that of land and sea mammals, from 7% to 10%. This criterion speaks in favour of our origin at the boundary of land and water. In anthropoids, the relative weight of the skeleton is more than in humans by about 40%.

Conclusion

In human nature there are many exclusive features, below we give a list of exclusives presented in the studies of different authors. Absolute anatomical and physiological exclusives exist only in humans: a special type of scalp hair growth; movable foot-spring; sweat glands of a special type, different from those of animals; increased sweating compared with other animals and qualitatively different physiologically than that of animals; location of sweat glands all over the body; cold sweat; a special blowhole device with holes pointing downwards; bell-like nostril cavities in their reflex tension; complex, stable bipedal locomotion including three elements: orthograde, bipedal locomotion and vertical posture. These elements are found only separately in ground animals; a special characteristic of the ear: a bell-like small cavity at the transition to the middle ear; turned out lips; the ability of abundant watering of the eyes; chin ledge; the ability to distinguish colours, combined with the ability to see forms of objects; the presence of specific melanocyte cells in the skin, which allow tanning under ultraviolet light; opposing thumbs, equal in strength to the other four fingers; a special group of muscles at the back of the hand (Saturn's Hill); the earlobe; abundant localised hair in the groin region; abundant localised hair in the underarm region.

Anatomical and physiological exclusives that distinguish man from the land mammals: scarcity of skeletal mass compared to terrestrial mammals' etiology; the composition of the blood, similar in elemental composition to seawater; the need for iodine; a combination of the oral cavity and nasal cavity into a single nasopharyngeal system; off season pairing; the lack of developed canines and diastema; a special menstrual cycle with the absence of severe heat-related bloody secretions, and the presence of menstruation, when fertilization is unlikely; the accumulation of fat of the type of marine mammals. Anatomical and physiological exclusives that distinguish man from apes and monkeys: the absence of supraorbital ridges; hypersexuality rather than the hyposexuality of apes; and the ability to tan.

Chapter 5

Establishment of humanity

Having considered the exclusive features of the human body and the issue of human nature one can talk about certain settings in search of human progenitors; one can define a more or less exact historical time when humanity was born, and the place where it appeared. There are three conditions or circumstances affecting the way in which man could have appeared: the first, human ancestors had to be universal creatures, able to live almost anywhere and to conduct a water-terrestrial life. The second condition is that the human body formed in a moderately warm and wet atmosphere. Thirdly, the basis upon which the human mind was synthesized could be a dissociated psyche of some creature. Presumably, it could have been an animal, in which the cranium had two relatively autonomous functioning brains. We have to find this creature and the source of this splitting.

Primate evolution

Looking for a human predecessor, let us look at initially the evolutionary history of the order of primate including the infraorder of simians. Monkeys, with the exception of tailless apes, emerged from pre-monkey prosimians relatively late. The existing prosimians include a few families of lemurs, tarsiers and the Lorisidae family. Monkey evolution began with the end of a terrible winter brought on by an asteroid strike around 66 Mya, blamed for eliminating the dinosaurs. They were the final products of specialization of certain types of prosimians. Part of the lemurs and tarsiers moved from hemihydrate lifestyle in a marine lifestyle (phenotype) whereas the other part continued the terrestrial evolution. The traditional (simian) approach in anthropology considers the prosimians as evolutionary precursors of monkeys, apes and humans. Currently, representatives of this taxon are found in Africa and Southeast Asia living in trees and active primarily at night or at dusk.

The transformation of prosimians in the Old World monkey was a not true progress but a mere continuation of specialization of life in the forest or savanna. The intellect of Old World monkeys (we will name them simply monkey, excluding from this concept apes and man) is not dissimilar from intellect of lemurs and much inferior to the intelligence of dolphins and great apes. Between the great apes and the monkeys like guenon and baboon, there is a chasm that cannot speak about their relationship. Including tailless apes and lower tailed monkeys in one order, based on the external similarity, is absolutely unjustified. Their origin and genesis are different. The evolutionary scheme of prosimians - the tailed monkeys – apes and man is the unjustified straightening of evolution. This is Linnaeus' view based on resemblance, without reliance on modern scientific data.

Primates are one of the most advanced groups of placental mammals, including monkeys, apes and humans. In 1758, Linnaeus singled out the order of primates and included in it humans, monkeys, prosimians, bats and sloths. For the defining characteristics of primates, Linnaeus took the existence of two mammary glands and five-toed limbs. Primates appeared about 65-75 Mya. The ancestors of primates lived in the trees of the rainforest and the lifestyle of most modern of primates is associated with trees. Accordingly, they are adapted to the three-dimensional environment. Currently, the order has more than 400 species.

Purgatorius is the first known primate that was found in the sediments of the Upper Cretaceous. It was small, about the size of a mouse, animal with primitive morphology, which lived on trees and ate insects. Possibly, it is an ancestor of the primates, although its taxonomic position remains uncertain. At the same time, there were other related groups of mammals: colugo, bats and the elephant shrew. At that time, there was a relatively rapid and massive increase in morphological differences between species due to adaptive changes and opening up of previously unavailable ecological space known as an evolutionary radiation.

About 55 Mya, there were adapoids, which were medium-sized tree primates, sized approximately as a cat, and widespread throughout Europe, Asia, Africa and North America. They fed on plants and insects and had a daytime or crepuscular lifestyle. The adapoids group was very diverse and in its formation occurred the second radiation of primates. Wortman (1903) suggested that adapoids were strongly related to anthropoids, now known as the Adapoid-Anthropoid Hypothesis. However, Gregory (1920) contrarily proposed that adapoid primates were more closely connected to lemurs and lorises, known as the adapoid-Strepsirrhine Hypothesis. Currently, this standpoint has far more supporters than the first hypothesis. Apparently, adapoids were the ancestors of modern lemurs.

Fossil lemurs in a modern form date from the middle of the Eocene. Ancient lemurs were found in Madagascar. In relatively modern times, the fauna of Madagascar included giant lemurs, such as *Megaladapis* that had the size of a large dog and weighed about 50 kg and are included in the family of *Lepilemuridae* that also include the small sportive lemur. In addition, there were *Archaeoindris* that had a height of 1.5 metres and weighed up to 200 kg, similar to the modern gorilla. These huge animals disappeared after the appearance of people who exterminated them on the island.

Omomyidae, about 50 Mya, were similar to adapoids but were smaller. They had a short muzzle, were nocturnal and fed on insects. It is assumed that *Omomyidae* were the ancestors of the tarsiers and the high primates and gave a second radiation of primates. Between 55 and 45 Mya, in the Indian sub-continent and China appeared Eosimiidae that in some essential features of tooth structure can be recognized as genuine ancient apes, although they retained many prosimian features.

Many ancient Old World monkeys, despite much apparent similarity to modern apes, were dead-end evolutionary branches and do not show an explicit relationship to modern apes. Some early Old World monkeys could reach a relatively large size, for example, *Anapithecus* could weigh 15-20 kg. It was believed that the hominoids and the Old World monkeys had a common ancestor and the separation between them could have occurred 28-30 Mya. The latest discovery of *Saadanius hijazensis* with dating 28-29 Mya, made in the west of Saudi Arabia, disproved this time of the divergence. The analysis of the length of the canine root and molar sizes determined that it was an adult male, weighing 15-20 kg, the size of a baboon. It lived in a warm, humid forest near the Tethys Sea where the Red Sea now is. It preceded the divergence of the lines of monkeys and hominoids (Zalmout, 2010). From this, the researchers concluded that the divergence between apes and Old World monkeys took place later than previously supposed, somewhere between 29 and 24 Mya.

Hominoids appeared in the Late Oligocene and have experienced their peak in the Miocene. Researchers have dug out a great many fossil species, of which the vast majority are dead-end forms. The most ancient and archaic is an extinct genus of primates Proconsul that existed 17-21 Mya in Africa. Four species have been classified to date, which differ mainly in size. The body weight for different types varies from 10 to 40 kg. They inhabited the tropical forests and combined the attributes of Old World monkey and apes. Characteristics of the proconsul, which are typical for an Old World monkey is a thin layer of enamel on the teeth, fragile physique, narrow chest, short forelimbs with a manner of movement in branches, leaning on all four paws. Hands and feet length are approximately equal and the limbs are not specialized. In addition, some species had tails. However, features of the tooth structure and the general proportions of the proconsul's skull, including a relatively increased brain cavity indicate belonging of proconsul to the apes. According to the traditional approach in anthropology, Proconsul has been included in a common ancestral lineage that leads to gibbons, apes and humans.

The history of the evolution of the extinct hominoid primate Oreopithecus from the late Miocene is very interesting. This primate, who lived on the territory of modern Italy, almost moved to an upright posture, even while walking, and stuck his hands behind his head hanging over the lower branches of trees. This is eloquently testified to by the structure of the Oreopithecus backbone, arms and legs. The skull is significantly shortened in the front office and brain, and though small in absolute terms, is fairly large relative to the facial features. Oreopithecus weighed around 35 kg. It possessed shearing crests on its molars that show a diet specializing in plant leaves. The canines are relatively small, that is one more sign of a hominid. However, this Oreopithecus was not a direct ancestor of humans; it was a wonderful example of parallel evolution. Oreopithecus lived in the swamp forests, with an absence of large predators. This allowed the monkeys to come down to the soil of the forest, where they lived. A cooling and drying climate led to the disappearance of the tropical rainforest. Swamps began to dry up and strong predators from nearby forests ended the evolution of these monkeys (Agust & Antón, 2002).

The evolutionary line leading to pongids and hominids in Africa and Europe was continued from Proconsul to the genus of Dryopithecus during the middle and late Miocene in Europe, Asia and Africa. A lower jaw with teeth and a humeral bone of Dryopithecus were found in France in 1856, by Larte. Other examples of Dryopithecus have been found in Hungary and China. Probably, this genus includes the common ancestor of gorillas, chimpanzees and humans. Dryopithecus were very similar to chimpanzees in structure and way of life, although they had a smaller brain and shorter arms. In length, it was 60 cm, and maybe had longer front legs, with which it moved from tree to tree like modern orangutans. It is also believed that Dryopithecus had a typical gregarious lifestyle. They lived on the trees and probably ate soft fruits and berries, as their molars are covered with a very thin layer of enamel. It is possible, according to the simian approach, that Dryopithecus became the progenitors of the first ancient hominids Australopithecines.

Thus, the traditional simian model of primate evolution consists of the following stages. Prosimians first appeared 70 Mya. Their likely ancestor is a small insectivorous animal, which transferred to life in the trees. Their descendants are the modern tarsiers. From these animals, about 40 Mya came real monkeys. From Propliopithecus, 35-25 Mya, came tailed monkeys similar to gibbons that lived in the tropics and separated *Pliopithecus* from the ancestors of Old World monkeys and apes. Approximately 22-20 Mya hominids separated from the branch of gibbon ancestors. *Afropithecus, Proconsuls* and *Ramapithecus* represented the other hominids. Then, 16-13 Mya, the branch Sivapithecus, the ancestor of orangutans living in South Asia, was separated. The other hominid, *Dryopithecus*, was the ancestors of pongids (human, chimpanzee and gorilla). From this, approximately 9-8 Mya, separated the branch Ouranopithecus (gorilla ancestors) and in the remaining 6 Mya evolved chimpanzees and the first ancient hominids Australopithecines, which are accurately described as two-legged apes, and as people with a monkey head. The complexity of the Australopithecines sub-tribe position among primates is that their mosaic structure combines features characteristic of both modern apes and humans. Australopithecines were discussed in the previous chapters.

Dolphin origin

The search for human ancestors and distant relatives is not confined to the rainforest or savannah but also is carried out in water. Assuming that the formation of human progenitors occurred in water, then moving to the boundary of two environments, and only then exiting onto the land. To find a sea mammal that could be the ancestor of man, among all the possibilities we should think of whales more than of monkeys and begin our analysis with the fossil and geologic history.

Having considered the exclusive features of the human body, the issue of human nature, one can discuss certain settings in search of human ancestors. We can also define, more or less, a historical time when humanity was born, and the place where it appeared. Human ancestors had to have been universal creatures, able to live almost anywhere and to conduct a water-terrestrial life. The landscape and the climate in which the human body formed were different from what we see today in most parts of Earth; certainly, it was not forest and savannah. The human body was formed in a moderately warm and wet atmosphere akin to a steam bath. The water with which pre sapiens had contact was maritime: salty, not a freshwater lake. The basis upon which the human mind was synthesized could be a dissociated psyche of some creature. Presumably, it could have been an animal in which the cranium had two relatively autonomous functioning brains. Therefore, we have to find this creature and the source of this splitting.

Solving the problem of human origin according to the above conditions, one can search for the most ancient human predecessors among the ancient delphinids (Ten 2011). Therefore, we need to consider the evolutionary history of toothed whales. Formation of the morphological type of toothed whales began about 60 Mya when there occurred the most extensive oceanic transgression, when along with the four ancient oceans formed the fifth ocean Tethys between the ancient continents of Gondwana and Laurasia. The relics of this ocean are the modern Mediterranean, Black and Caspian seas. For tens of millions of years, there was a drift of continents and Tethys was constantly changing shape. Around **60 Mya**, it was compressed to the size of the Indo-Atlantic Sea. The bottom of Tethys was a geologically active zone that warmed waters. It had gigantic geosynclines, around which arose the Alpine-Himalayan orogeny, creating the world's largest mountain system. There, about 60 Mya, on the rugged shores of Tethys, land animals currently unknown to science had begun to evolve into archaeocetes, the ancient cetaceans and the ancestors of the toothed whales such as dolphins and killer whales. The sea level rise encouraged the process as water flooded their habitats. Over time, they changed their habits from those of terrestrial animals to those of amphibians and then evolved into purely marine creatures. They had about 55 million years for the evolution that made them perfect marine predators, until about 5 Mya when they reached a form much like that of today.

There is much uncertainty, even at a simple first glance, about the origin of whales and the question of whether the toothed and baleen whales are of one or two different marine ancestors. The ancestors of archaeocetes are unknown, but it is believed that the ancestors of baleen and toothed whales were large ungulates. Traditional views on the cetacean evolution were that their closest relatives and ancestors were mesonychids, an extinct order of predatory ungulates that looked like wolves with hooves instead of claws. These animals had teeth of an unusual conical shape, similar to the teeth of cetaceans.

In addition, there is a concept based on genetic data that cetaceans are close relatives of cloven-hoofed animals, particularly hippos. In this case, one can see that some palaeontologists try to find the land ancestor of the whale on the based on the similarity of the geometric dimensions of the precursor and modern whales. However, the size of the land ancestor of the whale does not matter because over the course of 50 million years, a 160-tonne blue whale can evolve out of a shrew. In addition, the limbs of whales are, in their skeletal basis beneath the skin of the fins, five-fingered. These rays are very long and very elegant. They could not come from the hoof, because the hoof is also a formerly fingered limb, which was hit and bent by nature so as to lose the ability to unbend, and is ultimately a dead end of evolution.

It is important to note that the basis of all mammal and bird limbs were inherited from very distant amphibian ancestors. In the Jurassic time, there was no such diversity of extremities as nowadays. Everybody had one style: a many fingered limb with three or five rays, where the thumb is spaced apart from the other by an angle of 30 to 180 degrees, depending on the way of life. The limbs of mammals evolved from amphibian limbs that, in the evolution of various sorts of limbs, gave rise to pads, claws, hooves and fins. A hand was potentially very rich as a nucleus of evolution and its universal semi-finished product. In aquatic mammals, a hand is presented in its original form, but lengthened relative to body size and overgrown by muscles. Initially, it was a hand with membranes, then it developed into fins.

Palaeontologists for a long time marked a relationship between whales and ungulates; however, later was revealed the evolutionary link between the ungulates and primates (Spitsyn, 1989). Accordingly, there was a rapprochement between primates and cetaceans. Thus, whales, primates and cattle are close relatives that had a common ancestor that lived 65 Mya.

There is a fact of relative evolutionary youth of all ungulates, which began its existence later than dolphins. In addition, primates are older than dolphins and ungulates for example lemurs and tarsiers had already lived in the Paleocene Epoch 60 Mya, around 10 million years before the evolution of dolphins began. At that time, there was already a great variety of species of prosimians. They were inhabitants of tropical rainforests, coastal mangrove swamps and reed that were opposite to ungulates, which preferred a dry landscape. Therefore, despite the fact that it is hard to imagine a ten-centimetre lemur as an ancestor of a three-ton killer whale, one can assume that the dwarf forms of primates 60 Mya could be the ancestors of archaeocetes, given that they had 10 million years of evolution.

Possibly this was the scenario that described by Ten (2005). Before the catastrophe, when a huge asteroid collided with the Earth in North America about 66 Mya, lemurs and tarsiers were hiding in the rushes, not daring to go out for a long time on land or in the sea. The clouds of dust raised by the fall of the meteorite obscured the sun and a cold dusty night fell upon the Earth. The disaster destroyed huge cold-blooded predators and on dry land the possibility of the existence of life was reduced to a minimum. As a result, Tethys became the cradle of new life and a hotbed of evolution of new species; it was on the opposite side of the planet from the impact site, and between black clouds sunlight could sometimes appear, warming the Earth. In addition, Tethys was crossing a vast area of volcanic activity, through which the internal heat of the planet escaped to the surface. Thanks to geothermal springs, there was an ecosystem of protozoa, and shellfish that feed on protozoa, for which light was not needed at all.

Lemurs and tarsiers rushed into the warm, dark water. They ate, groping small molluscs in the dark with sensitive, long fingers and butchered them with nails. From nails instead of claws, came a sensitive, tenacious, flexible hand with a separate thumb. Lemurs and tarsiers had lost almost all methods of self-defence, because predators did not exist for a few million years. At the same time, the need for adaptation to new and severe conditions could trigger the growth of intelligence. The evolutionary new phenotype of semi-aquatic animals formed by obtaining and butchering molluscs in low light conditions and in warm water. It was this phenotype, that subsequently was inherited by people through some amazing transformations. Also, they were placental and that played a decisive role in their evolutionary competition with egg layers and marsupials.

It is important to note that prosimians are omnivores, as is man, as opposed to monkeys that are mostly vegetarians; the exception is a small predatory monkey Galago. Based on the criteria of universality, one can recognize that the nutrition of lemurs and tarsiers is closer to Homo than the monkey. Predation and vegetarianism are two dead-ends of evolution, each the result of specialization. Potentially further aromorphoses are unlikely in an animal that is specialized in a certain nutritional niche.

In conditions of the dust winter, their offspring were safe before childbirth and after birth they could hold their children in warm backwaters when the ground was covered by snow. Consequently, cubs grew up in water and gradually changed their appearance. In the warm water, fur started falling out, thinning, until it disappeared completely, and subcutaneous fat began to accumulate for thermoregulation. In water, body weight likely started to increase because the larger the mass, the less likely they were to freeze and the constant presence of water reduces the effect of gravity. Thus, the first steps toward becoming archaeocetes were made. As noted by the palaeontologist Mchedlidze (1976), one group of archaic terrestrial mammals moved into an aquatic environment at the end of the Cretaceous period during the transgression of Tethys. It was an animal, which had developed four limbs with five rays on each. Maybe it was a relative of the tarsier.

Tarsiers or lemurs, the former inhabitants of the coastal spaces, peninsulas and islands of Tethys, had developed movable limbs with long rays (fingers) and likely became the ancestors of toothed archaeocetes. By nutrition of wood snails, fruits, worms, and cane before the onset of the dust winter, they could easily move to the absorption of marine molluscs and algae. Long limbs allowed them to swim easily. It is very important that they did not have claws but had nails because in this case, limbs could much easier open shellfish. This hypothesis is partially confirmed by the existence of the medium-sized primates, Hapalemur, that live exclusively in tropical swamps on Madagascar. At the end of the last century, in Madagascar, a discovery of a giant fossil lemur was made of height of more than two metres. It lived in a cave, as did some ancient tribes of man, and such creatures could well turn into a dolphin over the course of 55 million years.

In the 1980s, in northern Pakistan, that about 50 Mya was a shallow part of the shoreline of Tethys, the expedition of Gingerich found among the remains of fish and shellfish two fragments of pelvic bone, apparently belonging to relatively large walking animals. Later, in the same place, the second of Gingerich's expeditions found a piece of a skull of a strange creature the size of a wolf. This unknown beast had characteristics reminiscent of the structure of the modern cetaceans' hearing system. It is important to note that this strange creature was found among the remains of marine and terrestrial animals. It was named Pakicetus that mean 'Pakistani whale' and was an extinct carnivorous semiaquatic archaeocetes, one of the oldest of the now known predecessors of modern whales, who lived about 48 Mya and probably adapted to the first attempts to search for food in the water by diving for fish. It had hind limbs, but the structure of the hearing system was similar to small whales and dolphins (Gingerich & Russell, 1981; Gingerich, et al. 1983). These findings of Gingerich's, which he humorously described as a whale with legs, support the hypothesis about the origin of cetaceans from lemurs inhabiting the border of the sea and land about 50-55 Mya.

Evolution of ancient whales

Conclusions about changes in the structure of the body and brain of animals in the process of evolution are made not only based on the fossil record but also on the study of embryos at different stages of their development. The idea of this approach was made by Haeckel and expressed in his phrase 'ontogeny recapitulates phylogeny' that as a biological theory shows that in developing from embryo to adult, animals go through periods resembling successive stages in the evolution of their distant ancestors. At present the following formulation biogenetic law is accepted: in ontogenesis a partial repetition is possible of certain features and processes that existed in the ontogeny of ancestral forms. For example, a man passes through stages that are similar to the different stages of development in embryos of fish, amphibians and reptiles. Cetaceans, at different embryonic stages, are also similar to the embryos of fish, amphibians and reptiles.

Rudiments of hind limbs in the embryo are formed in all modern cetaceans. However, they disappear at certain stages of embryogenesis. For example, the embryo of a dolphin of 75 days of age, has completely lost the rudiments of these limbs. It is obvious that the preservation of rudiments of hind limbs during the early stages of embryogenesis is none other than a reproduction of former features of the cetaceans' land ancestors. The forelegs are preserved in the five ray fins. In addition, the aquatic environment caused hydrostatic adaptations that helped to improve the buoyancy of the body, such as an increased the total area of the animal, the appearance of trabecular bones in the skeleton, the change of limbs to a horizontal orientation instead of vertical relative to the body (Mchedlidze, 1976).

In the early stages of development, a dolphin embryo has a round head and five-fingered limbs, it is therefore possible that the ancestor of these marine mammals was a land mammal. Pinnipedia are most likely descendants of ancient lemurs and tarsiers. They represent a transitional morphological type, although already firmly specialized, from coastal to marine animals. The lower monkeys are phylogenetically further from people than lemurs, representing the result of the development of lemurs and tarsiers in tropical forests in a later period (Ten, 2011).

We draw attention to the fact that the ancestors of dolphins had a mass comparable to the mass of human. The human skeleton developed a bone structure that became spongy, which does not meet the conditions of terrestrial habitats due to its weakness, but corresponds to the hydrosphere. Ancient whales' hind limbs had already become parallel to the body and limbs of this form are found in the embryos of modern dolphins. Another important hydrostatic adaptation was a movement of the breathing holes in the top of the head. The ancestors of dolphins had a blowhole located in the front, like lemurs. Moreover, the nasal bones hang over the breathing hole. Modern dolphins have a single hole located on top. Moving the blowhole caused a development of the triangular shape of a cetacean skull. Robust Australopithecus and modern gorillas have similar ridges that are one of the most archaic signs of anthropogenesis. Due to the ridges, the skulls of Australopithecus and gorillas have a form close to a pyramid. Similar skulls are found in Neanderthals and modern humans that suffer from microcephaly. Transition into the aquatic environment caused hydrodynamic adaptations like formation of the torpedo-like streamlined body shape, smooth external covering and reduction of fur (Mchedlidze, 1976).

In general, the dolphin embryo is remarkably similar to a small man with hands and a round head that, later in embryogenesis, is elongated

(Kuzmin, 1976). Both dolphin and man embryos are covered with soft fur lanugo, which then disappears. Mchedlidze (1976) noted that the Miocene dolphins had long humeral bones. One can say the ancestors of dolphins were even more like the human than the embryos of modern toothed whales. One can add to this the hind limbs discovered in archaeocetes and the ability of dolphins to swim vertically in water. They are the only marine animals that can move vertically in water, and there is not a good explanation of this phenomenon. It is possible that evolutionary memory motivates them.

Two features of the limbs of cetaceans attract particular attention: the stable preservation of a number of structures of terrestrial-like polydactyl limbs including nervous apparatus and maintenance of the ability to use them in certain behavioural acts. Metaphorically speaking, the central nervous system of modern dolphins still remembers the limbs of their ancestors (Vasilevskaya, 1984).

Acrodelfidy as possible ancestors of man

The rise of the water level of Tethys, lasting millions of years, led to many terrestrial-water animals including predecessors of whales losing their usual habitat, gradually moving to a water-terrestrial life style and then into a water life with periodic returning to the land for recreation and childbirth. Hydrodynamic and hydrostatic adaptations began, as mentioned above. One of the most important adaptations of animals using lung breath became a functional asymmetry of the cerebral hemispheres; two independent brain were housed in a single cranial cavity. That allows sleeping in one hemisphere and the other to stay awake. This allows the animal to not drown while sleeping.

Mchedlidze (1976) singled out three families among fossil archaeocetes of Miocene: *Squalodontidae, Acrodelphinidae* (proposed by Abel in 1905)

(Fossilworks, 2017, Carroll, 1988) and *Delphinidae* that have significant differences. Squalodontids are odontocetes like sharks with jaw and cheek teeth. It is considered that from squalodontids evolved most of the modern toothed whales, for example the sperm whale. The primary squalodontids tried to occupy the ecological niche already long occupied by numerous sharks and probably could not withstand the competition and so disappeared. Delphinidae had cheek teeth; the number of teeth had greatly increased in comparison with terrestrial mammals but the specialization of teeth had been lost. Molars and canines disappeared and so they lost the ability to chew food and crush shellfish. All teeth of delphinidae, as in modern toothed whales, are incisors. Modern dolphins are not able to chew food; they tear the prey and then swallow making them an effective marine predator. From delphinidae came modern dolphins and killer whales. Acrodelphinidae had incisors and molars and the total number of the teeth did not increase. Heterodontism of *Acrodelphinidae* shows that they chewed food and this could be done either in shallow water, taking a vertical position, or on the beach (Mchedlidze, 1976).

Flat teeth were used not only for chewing but also for chopping small molluscs. It was not an evolutionary innovation of *Acrodelphinidae*, as already in the Triassic period marine reptiles, Placodonts, had flat teeth for cracking the shells of molluscs and crustaceans. *Acrodelphinidae,* unlike *Delphinidae* and *Squalodontidae,* preferred shallow waters where there were a huge number of clams and had the opportunity to occasionally take a vertical position in order to chew food. They likely managed to preserve not only heterodont teeth but also limbs. They occupied an ecological niche, left by the delphinids and squalodontids which rushed to the sea depths.

Mchedlidze (1976) noted the findings of *Acrodelphinidae* with a round skull; one can assume that the later Acrodelphinidae ridge began to flatten. It played a positive role as long as the early *Acrodelphinidae* occupied the horizontal position during travel and leisure but the upright post was more beneficial when *Acrodelphinidae* fed upon molluscs in the shallows making the ridge redundant. In other words, evolutionarily, *Acrodelphinidae* in the process of adaptation to life in shallow water, began to come back to the original round skull shape while the blowhole moved forward. Acrodelphinidae had about 20 million years to return to the land from the sea and get comfortable.

The impetus for anthropomorphosis of marine mammals was the reverse process to the one that had led the ancestors of Acrodelphinidae into the sea some 60 million years earlier during the transgression of Tethys, when the sea flooded the land. Throughout the Neogene, Tethys had a regression when areas of submerged seafloor were exposed. Orogenesis in areas of prehistoric ocean contributed to the formation of a very favourable landscape for the development of higher forms of life on the territory of the future Eurasia.

The evolution of Acrodelphinidae began in the shallows of Tethys, which at the time of the beginning of anthropogenesis was two large bodies of water: the actual Tethys that was located on the territory of the Mediterranean Sea and Paratethys that was located from the Vienna Basin to the Aral Sea where what are now the Black and Caspian seas were its deepest depressions. Paratethys was periodically connected with Tethys via the Bosphorus. After losing connection with the Atlantic Ocean, the Mediterranean Sea enclosed from the north by the high Alps turned slowly into a giant warm bath with intensive evaporation, a unique environment for anthropogenesis.

During anthropogenesis, nowhere in the world were more favourable climatic conditions than in the Mediterranean. These conditions combined the two vital factors of evolution: epochal stability and gradual variability, wherein the second spurred life and saved it from stagnation. The evidence of the beginning of anthropogenesis around the Mediterranean is the findings of the most ancient creatures distantly resembling humans that were made there. It is known that *Anoiapithecus* is an extinct genus closely related to *Dryopithecus* that lived during the Miocene approximately 12 Mya, and *Pierolapithecus catalaunicus* which lived about 13 Mya in the territory of modern Spain. *Ankarapithecus* lived during the Late Miocene and its remains were found in central Turkey and the different *Ouranopithecus* from the Late Miocene were found in

Greece and Turkey.

We noted that from Delphinidae occurred modern dolphins and killer whales and from Squalodontids came modern toothed whales. Nevertheless, there appears one question: where are Acrodelphinidae? Why did this promising species suddenly disappear just when environmental conditions became more favourable for its development and the shallowing sea increase the food supply for them? Ten suggests that Acrodelphinidae did not disappear and they evolved as human ancestors. He believes that the start of anthropomorphosis of Acrodelphinidae began about 20 Mya in the southern part of the Tethys, whereas anthropogenesis began around 10 Mya. It is important to note that the last fossil Acrodelphinidae has an age of around 5 million years. Starting from this date, the lower boundary of anthropogenesis can be considered 6-7 Mya, consistent with the latest fossil finds of early hominids.

Ten (2005) suggests that different Pliocene pithecus are the result of development of Acrodelphinidae. Sivapithecus fossils, dated from 12.2 Mya, found in northern India may also the result of the evolution of Acrodelphinidae as in the Himalayan basin during the Miocene there existed similar conditions as in the western part of Tethys. These areas were not contiguous because in the Himalayan basin, instead of new drainage and flooding arose the Himalaya Mountains, and the connection with the western part of the former Tethys was interrupted. Possibly, Sivapithecus was a different, eastern line of hominization, geologically isolated from the west. Neotenic forms of Acrodelphinidae lived in a very elongated space, in the chain of shallow seas from Gibraltar to the Himalayan depression which existed prior to the alpine folding. For approximately 5 million years the geosyncline changed its relief splitting different groups of Acrodelphinidae from each other.

About 6 Mya, tectonic processes separated the Mediterranean basin from the Atlantic Ocean. The resulting large enclosed inland body of water **is** called Messina. In this closed basin was intensive salt accumulation. Thick layers of salt and gypsum were discovered during deep-sea drilling and echo sounding in different parts of the basin, mainly within the deep-sea trenches. This event is known in history as the Mediterranean Messinian salinity crisis. During this phase, the sea level decreased significantly by 1-3 kilometres. Some geologists suggest that the decrease was caused by chemical processes occurring simultaneously with the deposition of salts. On the dried bottom surface formed desert mud. The climatic conditions at the time elevated aridity in all parts of the Mediterranean basin. The Messina stage lasted a total of about 500 thousand years.

It is important to note that for at least half a million years prior to the Messinian salinity crisis existed favourable conditions for the ancestors of people with their water-earth existence. The climate was relatively stable, the salinity of the sea was not great and marine regression created estuaries, fiords, bays and islands. During the period of favourable conditions, 20-30 thousand generations of people could pass: sufficient for evolution. It is important to note that the age of the ancient hominids coincides with the period of the Messinian regression. That is, anthropogenesis could have started during the Messinian regression.

This hypothesis has partial confirmation by the finding, dated to approximately 5.7 Mya, just prior to the Messinian Salinity Crisis, of tetrapod footprints from the Trachilos locality in western Crete (Gierliński, 2017). The examination of the Trachilos footprints shows that the new species was bipedal, had plantigrade posture and the footprints have outlines that are distinct from modern great apes and look like those of hominins. The finding represents a previously unidentified late Miocene primate that evolved human-like foot anatomy. The age of the Trachilos footprints is roughly contemporary with Orrorin and somewhat younger than Sahelanthropus (see chapter 1). 'The non-divergent hallux and short lateral digits of the Trachilos tracks are absent in the foot skeleton of Ardipithecus ramidus' (Ibid.). That shows the existence of a species older than Ardi by a million years and yet having a much more human-like foot.

According to the traditional approach, the geographical range of pre-

Pleistocene hominins was East Africa (Ethiopia and Kenya) and Chad. Crete is some distance outside that area. Moreover, the authors note that Crete never had a direct connection to the southern Mediterranean coast and its late Miocene mammal fauna, including the new finding species, must have arrived from the north. The Trachilos footprints questioned the concept of African origin of humanity suggesting the idea that the homeland of humankind is the Mediterranean basin.

In 2017, in the Misliya cave (Israel), archaeologists discovered the ancient remains of modern man that includes an upper jawbone with eight teeth as well as complicated stone tools, smartly carved blades and sling projectiles that used to kill and butcher different animals. The fossil and tools, dated at between 177,000 and 194,000 years old, are much older than any previous Homo sapiens remains found in Eurasia (Hershkovitz, 2018).

According to the simianist approach, the earliest fossil recognized to *Homo sapiens* comes from eastern Africa and the emergence of the modern human is commonly placed at around 200 kya. The Misliya discovery breaks the existing simianist scenarios for the timing of the first known Homo sapiens because the finding indicates H. sapiens appeared more than 100,000 years earlier than they thought.

Practically at the same time, artefacts obtained from the archaeological site Jebel Irhoud, in Morocco between the Mediterranean and the Atlantic coasts, has also confirmed the view on modern human dispersal from the Mediterranean Basin. In Morocco, the researcher team uncovered human skull, face and jaw bones relating to five individuals identified as being early Homo sapiens and have been dated to between 280,000 and 350,000 years old. That indicates *H. sapiens* appeared more than 100,000 years earlier than the presumed origins of Homo sapiens in East Africa about 200,000 years ago (Hublin & Ben-Ncer, 2017).

In view of these two discoveries, there are two possible scenarios of human origin: the first occurrence of Homo sapiens in central Africa, not later than 500 kya. However, in this case, the numerous so-called predecessors and direct ancestors of *H. sapiens* such as *H. heidelbergensis* that lived between 600 kya and 200 kya, *H. erectus* that lived 1.9 Mya and many other archaic humans should be removed from the story of human evolution. As a result, researchers have to search for real human progenitors in Africa who lived 1 Mya and simultaneously they should erase most of anthropoid evolutionary trees, which have been built by the various academic schools. Otherwise, in the second scenario, one should accept that the Mediterranean Basin as the centre of human origin and there one should search for real human progenitors. Considering the inversion approach in this chapter, the author favours the second scenario.

The subsequent migration of distant human ancestors from the Mediterranean to equatorial Africa created new species. Depending on when they left the coastal habitat some 'migrants' from the seashore died out and they have been found in large numbers on the African continent by modern paleoanthropologists. Others were able to evolve into modern apes and the third group, who managed to stay in Messina for a sufficiently long time with a healthy diet and lifestyle, survived and spawned the races of modern people.

Frequently, modern anthropologists are trying to restore a process of anthropogenesis, according to fossil bones, without taking into account the bone pathologies resulting from the deficit of minerals and trace elements because they exclude the sea stage of the existence of human ancestors. An example of such bone pathology is the existence of the supraorbital (brow) ridges, specifically coarse thickening of the frontal lobes of the skull directly above the eyebrows of the massive Australopithecus, Neanderthals, as well as all modern ape anthropoids. That sometimes occurs in modern people in the case of atavism and some diseases such as acromegaly or cleidocranial dysplasia. However, the problem is that not all Homo sapiens ancestors had the supraorbital ridges. For example, Kenyanthropus platyops (flat-faced man) who existed 3.5 Mya and had a very flat face, as well as Cro-Magnon who lived 110,000 years ago, did not have brow ridges. On the other hand, 'younger species' such as Australopithecus robustus that lived around 1.5 Mya and Neanderthals of 40 Tya had noticeable superciliary arches.

Therefore, it is reasonable to pose the question: why do these large bone growths and thickening of the frontal part of the skull exist? A possible cause was a deficiency of vitamin D of hominids that had left the coastlands. This resulted in an X-shaped or O-shaped curvature of the legs; underdevelopment of the pelvis; pathological growth of the skull, which includes the growth of the frontal bone and appearing of the brow ridge. One can say that the skeleton of the Neanderthals, preserved by virtue of their evolutionary youth in comparison with the most ancient hominids, is an ideal picture of the long-time existence of vitamin D deficiency of the species. It has a stable, repetitive skeletal shape and a formed morphological type. They had potential to turn into apes, but they were prevented from doing this by the emergence of Homo sapiens that destroyed them and partially used them as genetic material. Their pathological human anatomy, which did not have time to transform into the normal anatomy of monkeys, developed over a very short period: from 20 to 40 thousand years: a moment in evolutionary history. Brow ridges are a pathology that quickly manifested and which is why it gives such a blurred picture in the retrospect of millions of years (Ten, 2013).

It can be assumed that the earlier departure of some pre-sapiens groups from the coast to the interior of the continents, both Africa and Asia, resulted in a significant change in nutrition from seafood to freshwater products and plant foods that led to a deficiency of minerals and microelements and created conditions to the degradation of the species and, especially, its brain structures.

When these species began to evolve into less sophisticated species, in the case of the chimpanzee for example, they had excess brain material, the specific mass of which progressively reduced because in the transition to a purely animal existence, an instinctive-reflex activity made a large brain an unnecessary luxury. In addition, the birth of cubs with large heads is very traumatic for both mothers and newborns and causes a large mortality of both; an animal with a large head has more difficulty jumping between tree branches. Therefore, evolution gradually brought everything into harmony. However, the contraction of the brain was faster than the decrease in the cerebral cranium because in the first case are soft tissues and in the second case are bone tissues. As a result, for millions of years, there was no brain pressure on the endocast as the image of the endocast is formed under the pressure of a growing brain but not a contractive brain. Accordingly, the chimpanzee's endocast is smooth and does not have traces of furrows on the surface, although it is not characteristic of the endocast of other apes (Shevchenko, 1971).

The essential feature of the endocast of fossil hominids is that the protrusions and depressions on them do not correspond to certain ridges and grooves of a brain. Unfortunately, this phenomenon has not yet been satisfactorily explained. The inversion theory of anthropogenesis explains the discrepancy between the endocast and the surface of the brains of fossil hominids by the pathology caused by avitaminosis D since in progressive rickets there is a pathological proliferation of the occipital and frontal lobes. Simultaneously, the brain had a pathological involution due to iodine deficiency. These two opposing processes lead to the fact that soft brain tissues shrank earlier. Hence, there is a discrepancy between the furrows and the gyri of the brains of the hominids and protuberances on the inner surface of the skull (Ten 2011).

As noted above, anthropogenesis actually started during the Messinian regression. However, after some time the salinity of the sea began to exceed the biologically acceptable norms and the sea gradually receded leaving a boggy marsh. Favourable biocenoses were preserved only in the mouths of rivers, notably in the Nile Delta. The slow drying of Tethys (later the Mediterranean Sea) caused a reduction of habitat and intensified the struggle for survival, in which only the strongest, or perhaps the smartest, survived and weaker members were killed or expelled. As the sea regressed, hominids were forced to concentrate in those places where beneficial conditions for development were maintained. However, it is important to note that from an evolutionary point of view, environmental stability is beneficial up to a certain point. Stability is essential to highly specialized organisms that could die at any change of environmental conditions but for creatures that have taken the path of universal development, instability is a stimulating factor. Therefore, even catastrophic changes were favourable to anthropogenesis.

The expelled pre-saliences tried to go to Africa because it was warm and free of the high mountains with their glaciers. There they managed to find tasty food, fresh water, tropical forests and there they lost rudiments of reason. There they, who were the runts of forming humanity, are found by modern physical anthropologists, thereby each time making a furore from anthropogenesis waste.

Another branch of the hominids went north to where Paratethys was located and where the natural conditions were different from Tethys. Into Paratethys flowed many of big rivers such as the Danube, Dniester, Dnieper, Don, Volga, Syr Darya and Uzboi and it never dried. In addition, above it warm and cold air masses intermingled, which led to the continuous formation of rain clouds, and so Paratethys was colder than Tethys and it was as the modern Mediterranean or the Black Sea, with a clear change of seasons.

Paedomorphosis

An important evolutionary significance may have paedomorphosis which is a form of evolutionary transformation of ontogeny based on heterochrony and which juvenile features of an ancestral organism are displaced to the adult forms of its descendants (Gould, 1985). The concept of paedomorphosis was proposed by Garstang (1922) who highlighted an ambiguity in the biogenetic law. He found that the biogenetic law, which required that adult stages of ancestors become visible in the juvenile stages of ontogenetic development, has contradictions because there is evidence of juvenile features of ancestors expressed in the adult forms of organisms.

He described its occurrence in salamanders, therefore, it has long been used mainly to refer to preservation of larval features in adulthood. In the presence of the life cycle of the larval stage, significantly different in structure, physiology and lifestyle from the adult organism, in some species in certain circumstances there is a retardation of morphogenesis of somatic structures while maintaining the normal rate of development of the reproductive system. As a result, the larvae become able to reproduce sexually. The retention of juvenile features in the adult animal is named neoteny.

The typical example of neoteny is axolotl (*Ambystoma mexicanum*) neotenic larvae of tailed amphibians, which due to hereditary lack hormone thyroid, remain in the larval stage (Armstrong & Malacinski, 1989). Young axolotls are unusual among amphibians because they reach adulthood without undertaking metamorphosis and their size is similar to adult individuals. The adult axolotl remains aquatic and gilled as an alternative to developing lungs and taking to land. When there is a gradual change in the conditions of existence or hormone injections, there is a metamorphosis of axolotl. Axolotls are able to produce offspring similar to themselves and the ratio of brain mass to an animal's total body mass (encephalization) higher than of Ambystoma whose neotenic form they are. Many biologists suppose that the greater encephalization of neotenic forms compared to the parent form is consistent pattern of paedomorphosis.

Paedomorphosis leads to complete loss of the adult stage of ontogenesis, a shortened life cycle and the stage that was previously larval becomes the final stage of development of the organism. Such paedomorphosis occurred, for example, in the evolution of some groups of tailed amphibians (proteus and sirens). From the standpoint of the evolution, neoteny is an important process, since it is a loss of rigid specialization, which is more characteristic of the final stages of development than for larval stages.

There are two main forms of paedomorphosis, progenesis and neoteny, which are adaptations to different environments. Progenesis takes place when the sexual development of an organism still in a juvenile period goes faster for instance some salamanders are able to reproduce in their larval life. Progenesis allows species to reproduce more rapidly and in great numbers and it can affect the evolution of new taxa, for the reason that it can reduce the developmental restrictions that arise in the development of organisms. Whereas the second form of paedomorphosis is neoteny that results in the retention of juvenile traits into adulthood. This is achieved by having sexual maturation take place despite the fact the individual is still in a pre-adult phase of phenotypic development. The larvae become sexually mature while retaining gills and some other larval features. Generally, neoteny causes species to reproduce more slowly and in moderate amounts (Gould 1977).

For changing the speed of development, sufficiently relatively small genetic changes initiate paedomorphosis so as not to cause infringements of the genetic balance (Ibid.). On the other hand, morphology acquired as a result of paedomorphosis is the juvenile ancestral morphology, generated by neoteny, and the organism retains the overall functional integration acquired in a previous evolutionary history of the ancestral form. These two factors determine paedomorphosis as a mechanism of rapid evolutionary changes where small genetic changes during paedomorphosis are accompanied by a disproportionately larger change of morphology, physiology and biochemistry of the organism while maintaining its integrity. Paedomorphosis is probably one of the reasons for the lack of transitional forms between major taxonomic groups.

Bolk (1926) found striking similarity between the appearance of man and ape embryos. In this regard, he put forward the hypothesis that man is a sexually mature ape embryo and that neoteny was an important factor in human evolution. Due to the developmental delay, evolution has the ability to turn in another direction without affecting adult features. The sprawling formation of all human organs, in comparison with those of the monkey, enable them to change depending on the environmental conditions at the embryonic level without waiting for maturation. A result of this, man has a relative weakness of the skeleton and muscles and a lack of hair. Other examples of preservation of anatomical features of the foetus (fœtalization) of modern man could be the proportion of the skull with a distinctive domed vault and the prevalence of the cranium over the jaw apparatus. Bolk believed that neotenic evolution of the human brain, which took place over a long time, gave man the highest capacity for learning and the formation of thought.

When the apes' reproductive system matures, growth of the body is usually discontinued; therefore, progenesis and retrogenesis should result in reduction in size of an adult individual. However, the entire evolutionary path from the earliest bipedal apes to Homo sapiens is characterised by a steady increase in the overall size of the body and the brain. It is also difficult to determine the type of human paedomorphosis since on one hand; it looks like retrogenesis because a retardation of the development begins in the embryonic period and impacts on the condition of somatic organs. However, on the other hand, the development of the reproductive system is significantly longer than that of the apes creating an abnormally long human childhood.

This is contrary to Bolk's concept that states that the juvenile stage occurred by progenesis or retrogenesis. We can see that man, in comparison with apes, has an elongated process of development and sexual maturation. Bolk's followers logically explain the origin of man based on the fact that man is neotenic form of some ape. However, their theoretical models cannot explain the difference between the brain of anthropoids and the human brain.

Montagu (1955) hypothesized that neoteny could be a pathway by which Homo sapiens could have progressed and so accelerated human evolution. He supposed that the childish *Australopithecus Africanus* could have a skull with a closer similarity to those of *Homo sapiens* than to those of the adult forms of their own species. The find has a deficiency of superciliary arches and sagittal crests, roundness of the skull and slimness of the skull bones together with an unusual shape of the teeth, comparative dimensions and form of the brain. It is important to note that neotenic forms of life are possible only in a certain group of animals. Species that appeared as a result of neoteny could be ground-based, but the neotenic original form could be either water (insects or frogs) or water-earth (axolotl) based. Terrestrial neotenic forms are unusual because the land animals cannot survive for a long time in childhood due to hazards around them. Dwelling on the border of two different environments gives water-terrestrial animals the possibility of effective escape from natural enemies, thus prolonging the period of childhood. Therefore, the Montagu hypothesis, in spite of its originality, is rather weak, as the proposed neotenic form of Australopithecines could not survive on the lower tier of the tropical forest and especially not in the savannah.

Possible human ancestor

Ten (2005) suggests that in the human evolution could be a period when development proceeded on the basis of paedomorphosis and the ancestor of man could be a neotenic form of Acrodelphinidae or as a second possibility, be a neotenic form of a delphinid. When the Tethys regression began, numerous groups of Acrodelphinidae (delphinid) were contained in some reservoirs that have no connection with the ocean as a result of the irregularity of the landscape. Evolution could offer these entrapped archaeocetes development only through paedomorphosis of the neotenic form. However, at the beginning of the regression of Tethys, Acrodelphinidae already had some preliminary adaptation to the shallows and so a reduction of the water area was a positive factor for evolution.

Mchedlidze (1976) noted that the ancient delphinid had been forced to return to the permanent shallow spawning grounds and development of dolphin embryos suggests that their predecessors had long arms, legs and round heads on necks that made them similar to people. In addition, biology suggests that delphinid could have the neotenic form; perhaps adults and cubs of the delphinid were different from each other in appearance, as adult and baby frogs and salamanders are different. Creeping to the shore, archaeocetes gave birth to creatures that were able to feed on the shore and in shallow water. Their children were similar to their ancestors which lived in two environments and about which we spoke earlier. Then, having gone through the metamorphosis, juveniles of the Acrodelphinidae (delphinid) forever departed into the sea.

Children of delphinids could be similar to people and could have been born in shallow water until evolution made dolphin birth possible in deep water. The fact is that, for all terrestrial mammals, newborns emerge from the uterus head forward (cephalic presentation). This provides a significant survival advantage at birth. Most predators, having the opportunity to give birth in a fairly comfortable environment, are not depleted as the mortality rate during childbirth is relatively low.

At the beginning of its evolution, babies of toothed whales came out of the uterus head forward and childbirth took place in this in shallow water because this provides rapid oxygenation of the newborns. However, due to the evolutionary movement into deep water, the head forward birth become impossible and evolution changed the motion of the newborn to reverse to tail first. Such an evolutionary adaptation as a change in position of the baby in the uterus, which allows birth in the deep waters, could happen over millions of years. Modern dolphins' babies come out from the womb tail first. Childbirth continues from fifteen minutes up to 2 hours and causes strong excitation of a flock. Females surround a birthing dolphin and her baby around the sides and sometimes below. The role of these midwives is to protect the baby from attack by sharks or adult dolphin males and to help the young if it is not be able to move by itself.

The tail forward birth of newborns contributes to their survival. However, these toothed whales, despite the safe environment, have a high infant mortality rate because dolphin babies often suffocate in the womb during the prolonged childbirth. It is a peculiar fee for new evolutionary adaptation and the ability to give birth in deep water.

According to ontogenesis, ancient delphinids had to have a neoteny form similar to man. Sexually mature embryos of archaeocetes had to be more similar to man than the embryo of modern dolphins, and so they were closer to the original form. It can be noted that the phenomenon of neoteny occurs if there are problems of survival of the adult animal's forms in the case of reducing the size of species, or the transition to it of unfavourable conditions of existence. In addition, there is a pattern that neotenic forms have an advantage in lower weight species and disadvantages in larger ones, compared to the ancestral forms.

Acrodelphinidae and people have a significantly lower weight in comparison with dolphins, so it is possible to consider such a neotenic delphinid ancient form. Thus, the neotenic form could have had an evolutionary advantage in comparison with the initial species (the dolphin).

When large groups of delphinids trapped within the boundaries of inland waters of Tethys, the neotenic archaeocetes form was more adapted to the changing conditions of the environment. It started as retardation, but went so far that it went into a steady paedomorphosis. Offspring of Acrodelphinidae frolicked in safe and warm creeks, engaged in sex, and were in no hurry to grow up, thus did not acquire a massive cigar-shaped dolphin body, which had no room to maneuver in shallow water. In this environment, they were more comfortable in their embryonic body with its hands, legs and head, which could twirl as desired (Ten, 2005).

There are also important additional arguments in favour of pedomorphosis, as the basis of human development. Anthropologists drew attention to the contradiction that exists between the traditional notions of the phylogeny of mankind and the indisputable data on ontogeny (Zaichenko *et al.* 2002). For example, a small degree of cranial base flexure and the forward angle of the foramen magnum (a large oval opening in the occipital bone of the skull that connects the cranial cavity with the vertebral canal), which can simultaneously be considered as a phylogenetically ancestral trait and as ontogenetically advanced. In other words, the anatomical features of man's cerebral cranium, reflecting adaptation to speech (cranial base flexure) and bipedalism (the foramen magnum) are a neotenic sign.

According to the traditional approach, phylogenetic-primitive characters are those features that could be observed in four-legged monkey human ancestors and which then disappeared in the course of development of the genus. The course of this historical progress must be reflected in the development of the embryo of modern man in accordance with the Haeckel biogenetic law to facilitate a partial repetition of certain features and processes that existed in the ontogeny of ancestral forms. For example, in the ontogenesis of dolphins the embryo passes through stages where it has rudiments of arms, legs and a round head, clearly hinting at the ancestral view. However, studies of Anisimova *et al.* (2002) show that, in the early stages of development of the modern human embryo, the skull has more apomorphic (advanced) Homo sapiens features than have the later stages. One can see that there is a contradiction between Haeckel's biogenetic law and the traditional theory of anthropogenesis using the hypothesis of gradual development from the non-human primate.

Within the framework of the inversion concept, this contradiction is removed if one takes into account that the origin of the upright position of the hind limbs occurred after the adaptation of the ancestors of the dolphins to the aquatic environment, when the legs occupied a position parallel to the body. Subsequently, the young neotenic archaeocetes originally relied on two legs, in small creeks where their childhood passed under the protection of their parents. After that, a metamorphosis took place in the delphinidae, which ceased to occur during further evolution. This phylogenetic history is reproduced in ontogenesis, in the development of the human embryo.

Inner speech as a phenomenon associated with communication between the two hemispheres of the brain, the two variants of 'I' (see Chapter 7), appeared before meaningful external verbal communication, i.e. before the appearance of *Homo sapiens*. Moreover, this inner speech formed a specific anatomy of the skull, reflecting the adaptation to speech activity and as a result, this phylogenetic past is repeated in ontogenesis, in the development of the human embryo. So the contradiction between the biogenetic law and the simian theory is resolved, the biogenetic law is not violated, but the basis of the traditional simian concept of anthropogenesis is in ruins.

Paedophilia and child sexuality

Zoologists testify that the phenomenon of child sexuality in the animal world is rare, but it is very developed among dolphins. The results of the study of sexual activity in cetaceans show that the accumulation of sexual experience begins in childhood. Male bottlenose dolphins can already have an erection at the age of 48 hours and babies try to copulate with adults at the age of several weeks; elements of sexual behaviour are observed between mother and baby. The stimulation of the mammary glands during feeding is probably arousing the mother; as a result, it takes the initiative and flirts with her cub (Caldwell & Caldwell, 1972).

In early human childhood, sexual behaviour is demonstrated within groups of children. There is also infantile sexuality associated with the carnal attraction to the mother. However, evolution always proceeds from necessity, in this case of sexual behaviour of the needs of procreation. The example is the natural heterosexual mating life of animals. The universality of child sexuality and its role in the formation of personality speaks of the enormous significance of this phenomenon in the phylogeny of humankind. Probably, child sexuality is the result of evolution by the mechanism of paedomorphosis and similar patterns of sexual behaviour in humans and dolphins could be another proof of the origin of man from the archaeocetes (Ten 2005).

Paedophilia, where an adult experience is an exclusive sexual attraction to prepubescent children, is a mirror of child sexuality; it is not only a criminal problem but also an unusual biological phenomenon. Until puberty, animals do not exhibit sexual interest between young and adults. There is childish sexuality in their world, but adult paedophilia is absent. Paedophilia can be considered as a psychiatric disorder and as a perversion of reversal. However, in recent times concepts such as normal and pathological human sexuality has occupied the attention of many sexologists and clinical psychologists because just as the definition of normality is historically and culturally built, the uncountable expressions of human sexuality, and varied searches for pleasure, confirm that sexuality in Homo sapiens has wide norms. Paedophilia is not only a perversion, but a violation of social norms. Society, at a certain stage of development, develops an offensive attitude to this relationship. History shows that it was practised in many ancient societies, in Greece and Rome for example, and has a long tradition that stretches through the history of primitive society. Therefore, one can pose a few questions: where and how did paedophilia get into the practice of primitive people? What peculiar properties of anthropogenesis could lead to the emergence of paedophilia in primitive societies?

Ten (2011) explains juvenile sexuality and paedophilia by paedomorphosis, which underlies the evolution of man. Evolution pushed the human ancestors to this because due to the change in natural conditions, neotenic individuals ceased to be modified in adult delphinids and there was a need for 'child' sex on behalf of prolongation of the genus. Thanks to this evolutionary adaptation, the neotenic form was morphologically fixed and began to develop independently, as a separate species, which in time led to the appearance of man. Children's sexuality and paedophilia in humans are social defects based on a biological evolutionary factor. This does not mean that these phenomena need not be combated by all possible means from pedagogical to punitive ones. On the contrary, it is necessary, because the social is the negation of the animal. Society began with this denial and is based upon it. Man is constantly obliged to make an effort not to succumb to the calls that return us to the animalistic state of being. The phylogenetic past looms over our present and constantly threatens it because between the human-social and the animal there is no and cannot be a secure relationship, this is a relationship of antagonism (Ibid). The transfer from the human state of being to the animalistic one can be made quickly and practically invisibly whereas the reverse movement from animals to human can require hundreds of generations.

Man playing and killing

The Dutch cultural historian Huizinga (1949) noted that if one analyses any human activity, it would seem like nothing more than a game. Moreover, human culture arises and develops in a game that is older than culture. He believed that all of the main features of the game are present in the equivalent games of animals, but the uniqueness of man is that the most important activities of human society are intertwined with the game where it oversteps the framework of biological activity. A game, rather than work, was a formative element of human culture; archaic society plays the same way as children and animals play. Before changing the environment, man made this transformation in his own imagination, in the sphere of the game. As noted by Berne (1964) social life mostly consists of games, and the lack of play leads to degenerative changes and death.

However, there is a difference between the play of man and play of animals. An adult man is much more a child than an adult wolf, lion or any other animal. All their life, people experience an inexplicable passion for the game, whereas for adult animals the game is virtually absent. Like all domestic cats, my cat Liza, when she was a kitten played all day because she had no problem getting food, so she had nothing to do. However, having matured and become a mother, she has largely lost interest in games and spends most of her life in hibernation or in anticipation of a favourable moment for stealing something from the kitchen table.

Ten (2011) believes that the game dependence of humanity says most likely that it has passed through paedomorphosis in its development so that infantilism is an indispensable component of human consciousness. In this aspect, it is possible to look quite differently at infantile behaviour in adults that is immaturity in development. The preservation in behaviour of features inherent in preceding stages is often explained in modern psychology by excessive comfort of upbringing and a lack of difficulties as a child. Within the framework of the inversion theory, there is a satisfactory way to define and explain unmotivated cruelty, which paradoxical as it may seem, often exists together with the infantilism of adult criminals. In my childhood, I had to observe how boys from good families hanged cats or threw them from a five-story building with laughter whilst being fully aware that this is not respectable behaviour. A great deal of evidence of unmotivated cruelty in children is found in different countries and across social groups. I want to emphasize the word 'unmotivated' when a child cannot explain the reasons for his actions or explains them ridiculously: "I killed just because", "it was interesting to watch" or "I was just bored". Mobile phones and the Internet have allowed viewing of examples of child cruelty by millions of people. Literature and art have paid attention to child cruelty; for example, Bradbury truthfully presented such stories: 'All summer in a day', 'The Veldt', 'The Small Assassin and 'Let's play poison'.

Inversion theory suggests that unmotivated cruelty is the legacy of paedomorphosis, because only in children is this phenomenon inherent in its purest form, when it is impossible to find motives for cruelty at all. Children of animals are also sometimes incredibly cruel. They are capable of killing when they do not want to eat, which adult predators rarely do. Apparently, this is also associated with the game and knowledge. Juvenile cruelty is a form of recognition of the world, its existential limits, extending far beyond the boundaries of the ordinary cruelty of adults. One can say that apparently unmotivated cruelty is motivated by cognition of the world, but only in the early child's development period. In adults, it manifests itself as an atavistic phenomenon associated with paedomorphosis which humanity passed in its phylogeny. While in adult animals, even the fiercest predators, it is not observed as a phenomenon, because animals in their phylogeny did not pass through paedomorphosis.

The existing cruelty among animals to representatives of their species is determined by intraspecific selection. For example, male polar bears that do not have natural enemies kill their own bear cubs, which are protected by the female bear, which cannot usually save more than one cub. The stronger bear cub survives, able to keep up with his mother. The male lion, having conquered another's pride, eats all the cubs. The relations within the herd of young male dolphins are very cruel, their games bringing the weak to death.

Ecological niche of a human progenitor

Currently, most anthropologists consider that the only ecological niche that could be taken by a human progenitor, which was a large animal with universal nutrition and which appeared on the planet at the time when diversity of species had already been established, was necrophagia (eating of corpses).

Primitive man, poorly armed, could not compete effectively with such predators as a machairodus or a wolf pack. In addition, it is unlikely that he was able to compete with numerous scavengers such as hyenas that were dangerous to humans. More importantly, the scavengers outpaced the humans in the speed of detection of feasting residues of large predators due to their better sense of smell. For the first people, who had just transited to a terrestrial existence, walking the ground was itself a fair amount of work. One can see the improbability of hunting and necrophagia in the first hominids.

A newborn's brain consumes 60% of the energy of the body, the adult human's brain consumes 25%, the brain of apes consumes 8% (Gibbons, 2007), that requires a very high-calorie food in ontogeny. Therefore, this raises the question of where our ancestors sourced high-calorie food if they were neither hunters nor necrophagic meat eaters. Perhaps the primary ecological niche of our ancestors was obtaining food that was hidden under a hard shell but not hunting or necrophagia. Dwelling in marine waters, our ancestors consumed mainly varying size marine molluscs (chiton). This ecological niche was environmentally exclusive to the human. From that time, the hominids left it and up to the modern day, the niche remains practically unoccupied.

There is no animal that can effectively obtain and uncover molluscs, for this an animal needs to have mobile and sensitive feet; considerable vertical elongation of the body; an ability to swim and dive; the ability of vertical bipedal locomotion in water; the ability to stay underwater for a relatively long time. In order to tear off affixed molluscs, the animal had to have a strong hand, with an accurate grasp and a spaced thumb to expose the molluscs; strong molars with flat surfaces to split small clams and break up the edges of larger shells and nails that can push through the gap in molluscs' covers.

There are several species whose bodies were also formed in a similar ecological niche. For example, the crab-eating macaque (*Macaca fascicularis*) which catches small crabs on the sand dunes or racoon (*Procyon lotor*) that is capable of catching and eating crayfish, using limbs very similar to those of a human. However, they are not able to catch and eat clams. Therefore, one cannot find creatures on the planet, except for man, whose body was formed in the conditions of this ecological niche.

In the Epipaleolithic (9000 years ago), the south of Europe was occupied by the Tardenoisian culture, named after the Fer-en-Tardenois site in France, where its centre was located. Traces of this culture are found everywhere in Europe, up to Germany and Poland. These people consisted of descendants of Cro-Magnon people with an admixture of Negroids. They widely started using trapezium microliths that are small stone flakes. Microliths served as arrowheads and inserts, which were inserted into the longitudinal grooves of various wooden and bone tools and fixed with tar. The discovery of microlithic technology was one of the genius inventions of humanity.

Usually, Tardenoisian's settlements were located along the banks of rivers and lakes, on the slopes of hills in fragile huts dug into sandy soil. Some settlements are found on seashores, marked by huge clumps of shells of edible molluscs. On the river Tahoe, in Portugal, 25km from its confluence into the Atlantic Ocean, is located a number of hills, composed of shells. One such hill, Cabezio d'Arrouda, reaches a length of 100m, a width of 60m and a height of 7m. In the layers of the hill, ash and coal layers, and less often bone items; bones of deer, bulls, wild boars, birds and fish were discovered (Mongait, 1973). People of the Tardenoisian culture consumed mainly clams; they extracted food from under the natural covers, from beneath the shells of molluscs, which were abundant in the shallow waters.

Similar food preferences are presented in the Mesolithic Asturian culture that originated in the coastal locations of Iberia. People of this culture used molluscs and snails as food. They left behind large clusters of shells of edible marine molluscs, among rudely studded pickaxes from pebble quartzite, used to collect shellfish on the coastal rocks, different ornaments as well as cutting tools made from shells of molluscs. Also, Asturian culture had a few of the most well-represented mythological traditions. Such as the entirely benevolent creature water fairy (*la xana*) who is a very beautiful young woman with long hair dressed in the traditional Asturian dress and who is associated with rivers and lakes as well as a dragon or snake (*el cuélebre*) that is an enormous serpent with wings and legs, and lives on the outskirts of a settled area in a cave. Of course, these creatures are similar to other mythical beings of many European regions and all of them have their roots in stories which are reflections of the experience of previous generations of human having a semi-aquatic existence.

Animals tend to be conservative in their food preferences, and primitive people are no exception. 10,000 years ago, the anthropogenesis of *Homo sapiens* had long been complete but Palaeolithic food preferences were preserved for a very long time. A large number of køkkenmødding (shellfish coverings) discarded by barbarians throughout the Mesolithic and Neolithic is additional evidence of the theory of the origin of man on the shores of warm seas, rather than in the forest or savannah. They were not engaged in hunting, but in obtaining food from under natural coverings, from the shells of molluscs.

Thus, in these creatures, the determining factor was the ability to extract soft tissue from hard covers. In the wild, molluscs are arranged in such a way that the protected tissue is the most valuable and digestible food, because otherwise there would be no need to hide it so well. For example, in mammals the brain is the most well protected tissue. One can assume that whilst passing from the water to the land surrounding the ancient sea, human ancestors occupied a similar ecological niche as they did in the water, consuming the most valuable and digestible food that was protected by solid covers. They started to consume brains and they were unique in the ability to extract brains from the bones. As a result, they managed to explore successfully new areas, due to the absence of competitors. Their feeding method was almost the same as in the waters, but instead of the shells of molluscs they split animal bones, particularly the skull.

The cave system of Zhoukoudian (near Beijing) yielded many archaeological discoveries in the 1920-30s, including one of the first specimens of *Homo erectus*. It was found layered with many metres of bone remains, dominated by the pierced skulls of hominids. It has been suggested that these skulls were the remains of meals where the main dish was brains. The heads of hominins and apes are compact and contain quite a lot of brain tissue. That is why they accumulated in vast quantities in the cave during the millions of years that it was inhabited. It can be assumed that the working day of hominids that acclimatised to the continental conditions was as follows. Coming out of the caves at sunrise, they observed the behaviour of birds and animals. Most of the large predators hunt at night and in the morning the remains of their meals are beginning to attract scavengers. They gnaw and peck bones, the commotion creating a great noise. Therefore, for human ancestors to find that place was not difficult.

As scavengers could not break the skull and get to the brain, this brain was left to hominids if rodents had not already eaten it. Fortunately, human predecessors often had time before that occurred, allowing them to develop as a species because brains are more nutritious than mussels and contain the full range of proteins and vitamins necessary for the construction of the human brain.

A fundamental change of habitat conditions posed problems for human ancestors. For example, how find the best way to split the bison's tibia to extract the marrow. Possibly, here began the use of tools. Nevertheless, the transition to land caused progress in the evolution of the species Homo but not all hominids. The degradation and then evolution to apes was experienced by those who had moved from the coast too early, unable to find a nutritional niche useful for the development of the brain due to a deficiency of essential micronutrients. Such was the fate of the overwhelming majority of hominid species and so they themselves become food for the more successful hominids, filling the Zhoukoudian cave with their pierced round skulls. Eating brain significantly increased levels of vitamin A, which in the form of retinol plays an important role in colour vision. Possibly, plentiful vitamin A in our ancestors' diet was a key cause of the ability to distinguish colours.

Birth of the races on Earth

In the animal world, colouring is generally determined by the need to blend into surroundings. Other reasons, such as a warning or threat are less common and this particular kind of colouring is typical of predators that have no competitors, for example, poisonous snakes. If we try to find any cause for pigmentation of a person, we are likely to turn to camouflage, but here we will fail. People live in the African jungle; therefore, they should be green. Nevertheless, they are black. The Vietnamese also live in the jungle, but they are yellow. Europeoids present a similar question. Actually, if we accept camouflage as the motivation of people's colouring, we all should be grey-green. In ancient times, the habitat of most people was the forest zone. The apes, which are considered to be human ancestors, also came out of the woods. Consequently, the modern theory of the origin of races reached a dead end. Therefore, to solve the problem of race it is necessary to find the landscapes that were identical in natural conditions, but different in the predominant colour. Moreover, the colour should be distinctive. In one case should prevail white, the other black, the third red and another yellow. About one hundred thousand years ago, people saw around themselves different colour landscapes. Their skin tone mimicked the predominant colours. Only in this way would race be obtained, distinguished by colour.

At first glance, it seems an impossible condition. However, if you look at the geographical map then many things become clear. Note how colours match the coastal cliffs. Coasts of warm countries have the same natural conditions, but contrasting colours that could be an answer for the question of why one kind of mammal, *Homo sapiens,* has a different colour. The white race emerged in southern Europe, near the Mediterranean. There all coastal cliffs are white and also there are white mountains. On the beaches white pebbles dominate. The yellow race appeared in Southeast Asia where the predominant colour of coasts is red-yellow. That is emphasized by names of rivers, for example, Mekong is 'red river' and Huang He is 'yellow river'. If you remove the anthropogenic landscape in Ceylon and the Indian subcontinent one can see a black shore.

Black people in Africa are newcomers, they came relatively recently. There are many archaeological and ethnographic evidence of this. Outside of a few thousand years, Negroid ancestry cannot be traced on the material culture in Africa. In the folklore of almost all tribes are the legends of another ancestral home for example folklore of Dogon (Mali, in Western Africa). The Negroid ancestral home is India where lies the real 'black continent'. Namely, Negroids created a civilization in the Indus Valley city of Harappa and Mohenjo-Daro. Coming at around two thousand BC, Aryans destroyed the Negro (Dravidian people) settlements, and pushed Negroid people far to the south in Ceylon and south of the Deccan, in the southern part of modern India.

Modern Dravidians are the descendants of ancient civilizations. The culture of Ancient Elam (pre-Iranian civilization) is ethnically Negro, as is the population of the Andaman Islands in the Indian Ocean. Aboriginals of Australia are a small race, which is a stable mix between black and yellow races. Negros expanded through Indochina, populated by yellow people with whom they mixed, before reaching Australia, then on to Polynesia. The fact that the black race formed off the coast of the Indian subcontinent, is also supported by the geography of settlements of Negroid people. In the Upper Paleolithic, archaeological finds place them in the middle of Eurasia, in Sungir near the modern Russian city of Vladimir.

The Russian Negros came not from Africa but from India. In Africa, they did not exist but in northern regions travelled and settled in competition with the Caucasian race. 'Black' anthropological material found in the

Upper Paleolithic site of Kostenky in the Russian Voronezh region and in the parking lot of Grimaldi in France. It seems that of the three major racial types, Negros were the first expansionists.

In the Upper Paleolithic is recorded their widest spread, almost all over the ecumene, while the settlements of white and yellow people are more localized. Perhaps this primitive imperialism and led the black race to its current situation; by the time of the beginning of civilization, the black race squandered its passionarity for expansion.

The black coastal landscape of Hindustan (Deccan Plateau) has geological origin; the plateau was made up of lava flows. At the end of the Cretaceous period, powerful trap volcanism reached two kilometres high, covering an area of nearly 500,000 square kilometres of the subcontinent. Trap volcanism gives another picture of sediments than one-time powerful volcanic eruptions. Trap volcanism leaves behind predominantly black and grey landscapes with smooth stones. When people came to the shore here, the coastal cliffs were black. Black basalt is characterized by curly forms; maybe lush hair well concealed their black body among foamy basalt on the shores of the Indian subcontinent.

The yellow race was formed by two factors: colour and conditions of environment. The coastal lowlands and estuaries of major rivers of Southeast Asia were yellow or orange-red colour, so the race was formed with a yellow and red complexion. Later, Redskins followed along the shore of the ocean to the north, across the Bering Strait to the territory of America, where Native American ethnic groups were created. In addition, the yellow race was formed in a calcium deficiency in contrast with the white race, which was formed under conditions of excess calcium in limestone massifs in Europe. As a result, the yellow race has a smaller skeleton in comparison to other racial types and higher flexibility which is associated with the development of cartilage instead of bone due to calcium deficiency (Ten 2011).

Ten (2011) believes that the formation of the yellow race occurred in constant contact with thermal mud which contains a lot of useful chemicals and biologically active elements and impacted on racial development. Contact with mud caused a particular type of hair: straight hair, tight, almost like bristles, and relatively sparse. Their flattened facial features are explained by the fact that, not being able to hide in the karst caves of coastal cliffs, as did the white race, or magma caves of Hindustan as black people did, yellow people were hiding in their mud lagoon marshes. This gave them long-term protection, provided that they managed to evade the many terrestrial and aquatic enemies.

The yellow race, often named as the Mongoloid race is one of the great races of mankind, widespread in North, East and South-East Asia. However, the yellow race at one time was called Mongoloid incorrectly, because the Mongol nation is large and powerful men unlike those whose ethnicity is mixed. The incorrect name of the race determined the path of discussions about the impact of landscape conditions to the formation of the race, assuming that the steppes and deserts formed Mongoloid race.

Meanwhile, apart from the Mongoloids in the steppes and deserts, also for a long time lived black and white people with non Mongoloid eye shape. Mongolians themselves became inhabitants of the Eurasian steppes rather late, about 10-15 thousand years after the end of race formation. Moreover, Mongolia was occupied by Caucasians until the 5th century BC. Scythian-Sarmatian cultures occupied a large part of modern China and Mongolia from the Yellow River in the South and all Eurasian steppes in the North.

A well-known anatomical feature of the yellow race is the specific shape of the eyes, which is why the yellow race is often called 'narrow-eyed'. There is a popular stereotype that this   feature formed in the steppe or semi-desert where there are a lot of dust and sand, which clogged eyes. However, the yellow race was not formed in the steppes, deserts and semi-deserts of Eurasia, where always lived people of white or black skin tone, as has been confirmed by archaeological findings.

The epicanthus is that special fold at the inner corner of the eye, to a greater or lesser extent covering the lachrymal tubercle and is an extension of the fold of the upper eyelid. It is one of the features characteristic of the Mongoloid race but it also exists in other races. There are a variety of factors that influence whether someone has epicanthic folds but there is no commonly accepted understanding how it was formed, so the hypothesis of Ten (2011) seems interesting.

Mud lagoons influenced the formation of eyelids, as thin eyelids were good protection against sticky mud. Getting out of the mud, which was hiding him from his enemies, early man cleaned his eyes with one hand movement on the stretched, thin skin of the eyelids. Long, thick eyelashes of a white man could create problems in the mud; they would have to be washed for a long time. But the semi-arid landscape contributed to eyelash development, so people with the Caucasoid type of eyelashes, thick and long, created additional protection against dusty winds.

The white race also formed in various forms with black and blond hair and with a steppe type of face, but these differences have a quite recent origin. It is important to note the general features of the white race: white skin and high, compared to other racial types, calcination of the body. Increased calcination can be explained by living among karst depositions where they consumed a lot of calcium in the water. The second reason was that in ancient times white people were inhabitants of the cliffs that were strewn by birds' nests. White people, of course, did not miss the opportunity to get this valuable protein food with high calcium content.

Separation of the races seems to have happened when man had already formed as a species. There is a great similarity between human races. There are no modern races without the chin whereas many species of hominids, for example, Pithecanthropus and Neanderthal did not have it. In all races, the supraorbital torus is missing and the largest molar tooth is the first rather than the second. The ratio of the width of the back of the head with the skull height and length of the skull base with the height of the skull are similar in all races and dramatically different from palaeoanthropes.

There is a reason to assume a very strong similarity in the manner of eruption of permanent teeth in the major races of humankind. As a rule, the second molars (Mr) erupt penultimate, that is before the third molars. In Neanderthals and Sinanthropus, the second molars M2 erupted before the canine (Schultz, 1960).

The chronological method, based on the rate of occurrence of genetic changes (the genetic clock), gives two options for calculating the age of emergence of racial differences: 800 generations, and 1500 generations (Khrisanfova & Perevozchikov, 2002). If it is considered that the duration of one generation is 20 years, it provides for a maximum of 30,000 years (the upper Paleolithic), that indicates a high rate of racial change.

Conclusion

Many human anatomical and physiological exclusives sharply distinguish man from apes, who have a completely different type of adaptation, characteristic of terrestrial animals. Therefore, one can hypothesize that man's distant ancestors were marine mammals. Specifically, man is an animal of marine aetiology, adapted to the terrestrial way of life as opposed to aquatic apes that are terrestrial aetiology animals adapted to a semi-aquatic life.

The possible landscape of anthropogenesis is the coastal border of water and land and not the tropical forest and savannah as supposed by supporters of the simian theory. Moreover, the specific nature of man is the absence of specificity.

Based on a number of conditions, human ancestors had to be universal creatures, able to live almost anywhere; to conduct a water-terrestrial life and the human body was formed in the moderately warm and wet atmosphere in salty water.

The place in the food chain is the most important of the causal factors of anthropogenesis. Human predecessors could not be effective hunters because they could not stand competition with species more adapted for hunting.

Scavengers have a specific phenotype and endocrinology, which is absent in humans. In this way, hunting, labour and necrophagy as the causal factors of anthropogenesis are more than controversial. Perhaps the primary ecological niche of human ancestors, when they occupied its early stages, was obtaining food hiding under a hard shell. Dwelling in marine waters, they consumed mainly chiton and this ecological niche was exclusive to the human predecessor. Thus, the determining factor was the ability to extract the soft tissue from hard covers.

Passing from the water to the land, human ancestors tried to occupy a similar ecological niche and they consumed the most valuable and digestible food that was protected by solid covers: brains.

Manifested as an instinct, labour as a causal factor of evolution leads species in such a narrow deadlock of specialization, where there is no possibility of the occurrence of consciousness.

In human evolution could be the period when the development proceeded on the type paedomorphosis and the ancestor of man could be a neotenic form of Acrodelphinidae. When the Tethys regression began, due to irregularity of the landscape, numerous groups of Acrodelphinidae were isolated in some reservoirs that had no connection with the ocean. Evolution could offer to land-locked archaeocetes only the paedomorphosis development through the neotenic form. At the beginning of the regression of Tethys, Acrodelphinidae already had some adaptation to the shallows so a reduction of the water area was a positive factor for evolution. Neoteny, in this case, meant that their cubs were able to feed onshore and in shallow water.

Chapter 6

The main human exclusive

The main human exclusive is the developed brain and therefore it has a separate, dedicated chapter in the book. The genesis of consciousness is the most stubborn sacrament about the human mind, due to its inherent conceptual problems. The question of the origins of consciousness has been addressed by many philosophers since antiquity, as they tried to understand what consciousness is and its nature. There are many theoretical approaches to understanding the nature of consciousness but few theories have been able to tackle fundamental questions such as: What is the nature of man's spiritual world? What is the adaptive benefit of consciousness? When did it evolve and do animals have it? One can attempt to find an answer to the question of how the consciousness of Homo sapiens formed. However, one should initially look at the history of the creation of psychogenesis theories.

Fundamental philosophical questions

Consciousness is the highest form of development of the psyche, characterized by a number of features. The main difference from animals: this form of the psyche is not limited to behaviour of the type of external stimulus-reaction with a reaction that is always expected, predetermined by instincts, or developed reflexively on the basis of external experience or training. In addition, the activity of consciousness can be completely unrelated to external stimuli, for example, creative people or autistics. An external stimulus, having entered consciousness, is processed by self-reflection. That is, the consciousness searches for answers to issues such as what is needed and what is wanted by someone who sent a signal of aggression or contact. The reaction of a normal person can be non-standard, unusual and unique, which usually does not happen in animals. If a dog, in response to the standard stimuli, answered the stimulus with an original reaction, the only one of its kind, we would suspect that consciousness awoke in it.

Generally, consciousness is understood as the highest level of man's mental structure, where man separates himself from his environment and reflects in the form of mental images, which work as supervisors of goal-oriented activity (Spirkin, 1983). The consciousness is a product of history and the result of the practical and cognitive activity of a thousand generations. It has a social history and a natural pre-history, the development of its biological fundamentals took the form of the evolution of intellectual activity in animals. The psyche is the possession of the entire animal kingdom. Consciousness is inherent only in man, and even then not in every state. It is absent in a newborn, in some types of the mentally ill, in a person in a state of sleep. Even with a fully developed person, the main stock of experience is stored outside the immediate control of consciousness, a person experiences all the physiologically accessible effects, but not all of them become a fact of consciousness (Ibid.).

Consciousness refers to the ability to reflect reality, or transpose an object into subjective content of a person's awareness, as well as specific socio-psychological mechanisms and forms of such reflection at different levels. Consciousness is understood not just as mental reflection, but also as the highest form of purposeful mental reflection of reality by a socially developed person.

When psychologists and philosophers talk about consciousness, one of the main problems is confusion in the definition of consciousness. There are hundreds of contradictory definitions of this concept, perhaps reflecting a number of unresolved problems in psychology. It is interesting that the lack of good terms often illustrates the exceptional complexity of the object of study. Theories of consciousness come from philosophy, cognitive science, biology and religion. The definitions of consciousness are usually given in discourse of origins, and not in the essential nature of the consciousness. This is an easy way of bypassing the logic of definitions. It is like defining a tree by describing how it grows, from what kind of seed and under what conditions. All the materialistic and evolutionary definitions of consciousness use the concept of reflection. Consciousness is an active reflection of reality; inside itself it processes external impulses. As a matter of fact, this is the answer not to the question of what consciousness is, but rather how it arises as reflection: from simple forms such as chemotaxis to increasingly complex ones. However, the problem is that the path of complication is well traced from chemotaxis to the reflexes of higher animals and further research faces a cognitive barrier. For several centuries, no attempts of scientists have been able to traverse the gulf between animals and man. No one can even understand what this gulf is, where it came from, and yet within it lays the secret of consciousness.

The origins of consciousness have been addressed by many philosophers since ancient times, trying to comprehend what consciousness is and its nature. Therefore, we will place the study of the origin of consciousness in a historical perspective.

Descartes

Of course, one can list ancient thinkers, in whose works the problem of the source and nature of human consciousness was the main theme. However, one can assume that consciousness as an object of research was introduced into science by René Descartes, who was born in 1596 in France and was both a mathematician and philosopher. Before him, the subject of analysis was not consciousness but the spirit and the soul and this knowledge was esoteric discourse.

From his standpoint, consciousness was non-spatial substance; open only to the subject contemplating it and engaged in the process of cognition of reality. Descartes described all internal conscious phenomena of psyche as thinking, contrasting them to emotions. He tried to approach the study of emotions as appropriate to the study of natural phenomena, rejecting religious and moral views and prejudices. Descartes tried to extend the mechanical principle of an explanation of life of the organism. For that, he introduced into science the concept of reflex, which played an important role in modern physiology of the nervous system.

Descartes fully separated a body and soul, and recognized the existence of matter that is an extended substance and the soul, which is the thinking substance but not extended. Both of these substances are defined by different attributes and are mutually independent ideas. He equates consciousness with thinking, but by the term thinking he understood everything that takes place in us so that we perceive it directly ourselves. Therefore, not only to understand, desire, and imagine, but also to feel means the same thing here, all being considered as thinking.

He defines spirit as everything that happens in us so that we ourselves directly perceive it in ourselves; in other words, Descartes introduces the principle of reflection of consciousness in itself as a self-contained internal world. Descartes brought thinking to the fore and, thereby, made its examination possible in secular science while the soul was relegated to second place and stayed in theology as divine substance.

Reflection is thinking about oneself, and according to Descartes, it is the most convincing evidence of existence itself. He identified thinking with the mind and derived the fact of his existence from his thinking essence. In this case, Descartes determined his principle as 'Cogito Ergo Sum', that means I think, therefore, I am. He supposed that this principle cannot be doubted, because any effort to doubt 'Cogito Ergo Sum' confirms the fact of his existence, as doubting itself suggest thinking, and a person has to exist in order to think. He wrote (Descartes, 1911, p. 9):

> *I do not now admit anything, which is not necessarily true: to speak accurately I am not more than a thing, which thinks that is to say a mind or a soul or an understanding, or a reason, which are terms whose significance was formerly unknown to me. I am, however, a real thing and really exist; but what thing? I have answered: a thing which thinks.*

The development of the science can be represented in the form of successive attempts to find the origins of consciousness and to determine consciousness through its dialectical opposition; therefore, the subject of research in different paradigms can be imagined as a set of dichotomies of consciousness and its dialectical antithesis. The first dichotomy was a psychophysical problem (the dichotomy of consciousness and matter) the problem of the relationship of the psychic with the matter that Descartes posed.

The differentiation of the concepts of thinking and consciousness in philosophy only came in the late 19th century and as a result, a weak substantiality of the concept of thinking as a process was identified, whereas consciousness contains something more than the process and it this gives a degree of substantiality. Of course, strictly speaking, in the mind there are no pure substantives, so the difference can be considered as semantic.

The Descartes doctrine had become the basis for the epistemological views of English philosopher John Locke that divided human experience on the internal concerning the activities of the human mind, and external oriented to the outside world. Defining consciousness as 'the perception of what passes in a Man's own mind' (Locke, 1979), he actually offered a method of investigating consciousness that later was called as introspection. Locke believed that there are two sources of human knowledge, objects of the external world and the activities of one's own mind. A man directs his external senses upon the external world and so he receives impressions about outward things. Activities of the human mind, in which Locke counted thinking, doubt, faith, discourse, knowledge and desire are cognized by using special, internal feelings – reflection, the observation that the mind places upon its own activities (Ibid).

Introspection

At the end of the nineteenth century, introspection was the main method of studying consciousness as researchers thought that the process of consciousness is apparent only to the subject of this consciousness, and is obscured to an outside observer. Thus, no one else can understand the state of another man's mind as he does his own.

The main apologist of introspection was German physiologist Wilhelm Wundt, who first raised the question of the necessity of studying psychological phenomena through sociogenesis due to the fact that there are no abstract individuals who have not passed through the crucible of social genesis. Introspection attracted masses of supporters and introspective reports were published in scientific journals in large numbers. Laboratory tests were carried out under strictly controlled conditions to obtain agreement of the results from different examinees. Examinees were presented some visual or auditory stimuli such as images of objects, words or phrases; they were expected to perceive them, compare them with each other and inform researchers about associations which arose.

However, with the expansion of introspective studies, problems were found. The results did not coincide not only between the various authors, but sometimes even between studies of the same author when working with different examinees and an unpleasant consequence of this was accumulating contradictions in the results. Later, systematic use of introspection came to detect non-sensuous elements of consciousness. Finally, unconscious causes of certain phenomena of consciousness became apparent. Researchers began to find such aspects of consciousness, which cannot be decomposed into separate sensations, nor presented as their sum. For example, if an experimenter were to take a melody and transfer it to another tonality; each sound will change it, but the melody in this case remains and easily guessed by a test subject. Hence, separate sounds do not define the melody but rather it is defined by some special quality, associated with the relationship between the sounds. This quality of the whole structure was named Gestalt and became a key element of Gestalt psychology developed in the Berlin School of Experimental Psychology. Gestaltists did not like the dismemberment of the subject of psychology into individual behavioural acts as they did not see this as a way to understand consciousness as a whole. They are characterized by an integrated approach outlined as far back as the time of German philosopher Leibniz. Gestalt psychology is characterized by the use of categories that underscore the integrity and indivisibility of perception. It spread widely concepts such as insight, short circuit, transposition and isomorphism. The last theoretical concept is the general idea of the matter of the origin of consciousness and is the structural similarity or identity between areas of nerve excitation in the cerebral cortex and the way in which they are represented in consciousness. In other words, it expresses the idea that Gestalt qualities such as shape, configuration, organization and structure of physiological processes in the brain that underlie subjective cognitive experience are identical to Gestalt qualities of shape, configuration, organization and structure of the subjective experiences of the subject.

In connection with the development of physiology and anatomy in the nineteenth century, the aforementioned psychophysical problem was, for many philosophers and psychologists, transformed into a psychophysiological one, which considers the relationship of the psyche to the processes that occur in the nerve substrate. The second dichotomy became a psychophysiological problem of the consciousness of man and the anatomical vessel in which it is carried.

Objective psychology

In the late 19th century, the lack of prospects of introspective methods in experimental psychology led to an understanding of the importance of developing an objective method of study of the phenomena of consciousness. One can say that the statement of the objective method was the start of scientific psychology as an independent science. At the source of objective psychology was a Russian physiologist Sechenov who set about the task of creating psychology as a science, based on an objective method, which he said is a native sister of physiology.

He found that psychological phenomena have a reflex nature and gave an opportunity to understand the psyche as a system of activity, formed under the influence of the outside world. Reflexes underlying mental phenomena are reflective neural mechanisms and the process begins with perception of a stimulus, continues in neural processes of the brain cortex and ends with a response of the body, primarily by muscle movements. He assumed that every sort of the central nervous system's activity is a reflex derivative and can be visualised as three basic bonds: the stimulation of sensory receptors, the action of a reflex centre and the excitation of the effector systems.

'When the child laughs seeing toys, when Garibaldi smiles when being driven by excessive love of his homeland, when a girl trembles at the first thoughts of love, when Newton creates universal laws and writes them down on paper, everywhere the final act is a muscular movement' (Sechenov, 1866). Sechenov believed that all acts of conscious and unconscious life originate in reflexes and he denied the separation of reflex and conscious activity. Sechenov put forward the promising idea about the act of consciousness, as 'an unfinished reflex', the body's response to an external factor is not recounted to physical action. The study did not focus on the definition but on an understanding of how consciousness arises and thus reformulated the research paradigm.

Ukhtomsky, who was a pupil of one of Sechenov's pupils, when analysing the action of a reflex centre developed the concept of a dominant. The idea of a dominant consists of hundreds of separate body mechanisms, independent from each other, but all subordinate to the brain. An awake person constantly switches from one movement to another and the work of different mechanisms are replaced kaleidoscopically. Therefore, the central nervous system requires a locking mechanism of great power because in order to function, the brain has to block all such influences. It was found that inhibition occupies a significant part of the brain's energy. Launching one mechanism, the brain is simultaneously blocking another and it takes more energy than the directly performed work. Inhibition is an expensive excitation.

One can say that the creation of prerequisites of the work is harder than the work itself. It is difficult to imagine such waste in nature. The question arises, how can we ever do anything? English physiologist Sherrington, who discovered Sherrington's funnel effect (final common pathway) as a physiological principle, offered that the most excited nervous centre takes the initiative. In 1920s, Ukhtomsky found another explanation of this problem. 'The dominating centre reinforces its own excitation through collateral impulses. Simultaneously, with the excitation development in the dominant, it inhibits other effective reflexes in the final common pathway' (Ukhtomsky, 2002). In addition, a dominant has a capability to exploit nonessential stimuli for its own purposes, which is quintessential of any dominant for the period of the height of its activity. He noted that this inertia consists in the fact that an already caused dominant is able to persist for some time in the centres and reinforce both its excitatory and inhibitory elements by various and distantly related stimuli. That's why incoming stimuli, which should simultaneously cause many sorts of reflexes, do not destroy the body, but promote the effect of the single reflex arc that at the moment is dominant (Ibid). This phenomenon was called the dominant by Ukhtomsky.

He wrote that the dominant effect, working as the energy expropriator of many focuses of excitation in order to direct all energy to a narrow channel of excitation. However, the influx of energy from large-scale inhibition, converted into one dominant excitation, necessitates new inhibition. Finally, it obtains coincidence in phase with its own dominant force and the inhibition caused by itself. This condition is called parabiosis and was discovered by a teacher of Ukhtomsky, Vvedensky. Physiologically, it looks like persistent inactivity agitation (such as a convulsion), when the tissue loses the conduction ability of nerve impulses.

The concept of dominance can give an explanation of many aspects of the human psyche such as different modalities of perception and emotions. During the time in which dominance continues, it gives the integrity of perception that is not a passive action; it is subordinated to the actual objective through a dominant. Dominance is only possible at the beginning of an excitation. The more stimulation, the faster it will cause an adequate inhibition. So, evolution has created a survival instinct that balances excitation and inhibition, avoiding situations of parabiosis and which works not only to avoid situations that threaten a total loss of function, but rejects actions that require aimless expenditure of vital energy. For example, self-preservation instincts, acting by reflex, teach the dog during maturation to ignore small birds, which it enthusiastically chased as a puppy.

The doctrine of reflexes

Ivan Petrovich Pavlov

In first decades of the 20th century when the dominance theory was being created, the main research directions of physiology studied the correspondence between external stimuli and bodily reactions and the correspondence between brain areas and peripheral effects. The first of these directions was actively developed by Nobel laureate Ivan Petrovich Pavlov, whose fundamental concepts of learning by association are broadly recognized in mass culture, for example: 'when you call my name I salivate like a Pavlov dog' sing the Rolling Stones.

Pavlov (1955) revealed that unconditioned reflexes, which are inborn and spontaneous, need no learning, and as a rule are similar for all members of a species. In the process of individual experience or learning, conditioned reflexes are formed that may differ a great deal between individual members of a species. They consist of a conditioned stimulus, such as the bell in the Pavlov's experiment, producing a conditioned response such as a dog's salivation.

An important principle of the reflex activity is the principle of reinforcement of reflex activity that achieves results. The reinforcement of reflexes is implemented by the effects of actions themselves using a feedback mechanism; when any reflex actuates an appropriate effector apparatus, such a muscle, the pulses occurring in the operation of the muscle return to the central unit of the reflex. These indicate not only about the action of the organ, but also about its results, making it possible to adjust the continuing action and ensure the adequate implementation of a goal.

Pavlov's study showed that all conditioned reflex activities, by their biological nature, are signals. This means that based on the formation of temporary connections, numerous signals both from the external environment and internal to the body (conditioned stimuli) are precursors of the upcoming activity, essential for the body's unconditioned reflex actions such as digestion or some sexual functions.

Preventive action of a signal in advance (anticipatory reflection) prepares an organism for the forthcoming act of satisfying some need. Incoming signals to the brain, caused by objects and phenomena acting on the sensory organs, were called the first signalling system by Pavlov. However, a man, unlike other animals, also has an inherent second signal system, related to human speech. According to Pavlov, the word is a signal for the first signal system. The second signal system was formed in humans in connection with the social lifestyle in which it is a means of communication. Speech and writing are not only auditory or visual stimuli; they carry certain information about an object or phenomenon. After the formation of a conditioned reflex to a certain stimulus in a man, it is easily reproduced without reinforcement, if this stimulus is verbally expressed.

Introducing the concepts of the first and second signal systems for distinguishing between the instinctive-reflex mode of action and the signs of the human language, Pavlov drew attention to the antagonism of these two. It turned out that the second signal system constantly has an overwhelming and depressing effect on the first. Graphically, they should not be represented as two tubes, conveniently nested one inside the other, but as two cones with points facing each other. Thus, these two systems can be depicted as a one-sheeted hyperboloid of rotation wherein the throat-shaped ellipse there can be an inter-transition of two signal systems. Pavlov came to the conclusion that consciousness cannot be determined by objective methods of reflexology, since it does not have adequate tools for such studies. Consciousness is a thing in itself, an intelligible object existing regardless of our perception. In contrast, reflexes are sensually perceived phenomena. Pavlov's conclusion can be considered as the methodological basis of the new approach of psychogenesis.

The development of Pavlov's ideas led to the development of a fundamentally new paradigm for the study of the emergence of consciousness that is based on the dialectical logic of the negation of the negation with sublation as per the German verb *aufheben*, which means to cancel, to preserve and to collect. Using the Hegelian dialectic, one can say that the first negation is a transition of the initial state of the system (reflex and instinct) to its opposite (consciousness). However, this opposition contains one-sidedness, incompleteness of manifestation in the development of the essence of the object and the impossibility of existence. This incompleteness and one-sidedness must be overcome by the second negation, which is the negation of the negation, this is the removal of the opposites (the initial stage and its negation) in their new qualitative unity, and their dialectical synthesis. Dialectical synthesis means establishing a higher quality system than was in the first two stages. This is not a simple sum of one and the other, original and opposite (reflex and consciousness), but something new that arose as a result of resolving the contradictions of the previous stages of development. Dialectic synthesis also ensures the preservation of the content components, the acquisition of new links and the development of new integrity. As a result, the system passes to a new, more advanced stage of development. As applied to anthropogenesis, the dialectical approach means impossibility through constant complication of the reflexes to achieve the consciousness of modern man. The latter can be achieved only as noted by the development of the development of inversion principles in the genesis of the human psyche. Consciousness can be formed only by the development of inversion principles in the genesis of the human psyche (the inversion theory is presented in Chapter 7).

The doctrine of reflexes combined physiology and behavioural problems. Researchers represented consciousness in various forms in the early studies of reflexes, for example, in the form of an unfinished reflex, a complex reflex or as a sum of reflexes. In the end, the researchers came to the conclusion that reflex and conscious actions are not only dissimilar but in opposition. Consciousness does not arise where more and more complex reflexes are developed, but rather where there is a rejection of the instinctive, reflex, animal mode of action. Complex reflexes, such as in bees, are fixed in complex instincts and not only does this not contribute to the development of consciousness, but indeed makes the development of consciousness impossible.

Köhler

The early 20th century was a time of intellectual debate between introspectionists and supporters of the objective method. They made many discoveries about the methods of a functioning consciousness, but came no closer to the issue that is the origin of such consciousness. Darwin's theory of natural selection triggered experimental work with animals as a new object of research. Instead of asking what consciousness is, more and more frequently a question is raised of how consciousness arose. At first, it seemed that the problem was simple and it would be easy to find the answer by examining the animal psyche. Development of the concept of thinking was strongly influenced by Köhler's experiments with apes on Tenerife, which caused a sensation in both science and society in the 1920s. These experiments tried to prove that chimps are different from other animals and they have thinking and language and consisted of placing chimps in an enclosed area and presenting them with bananas that were out of reach. In one experiment, Köhler placed bananas outside a cage and two bamboo sticks inside the cage that were not long enough to reach the bananas. Therefore, there was only one-way to reach the bananas: to connect the sticks together. Moreover, the researcher demonstrated to the apes by putting his fingers into the end of one of the sticks but this did not help to animals to find the solution. After some contemplation, the ape made many attempts that did not work. They began to get nervous and dropped to the floor. Some chimps abandoned this as a futile exercise and switched to something else or sometimes just gave up and started sitting (Hothersall, 2003). After that, occasionally, the ape returned to the task and after some contemplation, a miracle happened. The ape put the two sticks together and formed a stick long enough to reach the bananas outside his cage. The conclusion that Köhler made was that it spent its free time thinking how to solve the problem and found the solution.

Pavlov, however, noted that Köhler did not see anything that the apes really showed him. The desire to make a psychological difference between a monkey and a dog in the associative process is a hidden desire of psychologists to move away from the clear solution to the problem, and to make it mysterious and special. He added that in this harmful quest to get away from the truth Köhler used such empty concepts as a monkey going away, thinking in a human manner and deciding the matter. It is known very well that the dog often cannot solve a problem but if it is allowed to rest, then the dog can solve it. Pavlov argued that this was not demonstration of thinking, but rather that due to fatigue inhibition appeared. This inhibition blurs, hinders and destroys previous memories of the task. This is an associative process, using the process of analysis using analysers, with participation of the inhibition process in order to differentiate something that does not meet the conditions. Consequently, we cannot say that the monkey has some kind of intelligence that dogs do not have, and which brings him closer to man, and instead use only associative processes.

The statements of Pavlov were based on the theory of the dominant, developing the doctrine of parabiosis by Ukhtomsky and Vvedensky. In Köhler's experiment, an ape departs from the trough due to conservation of vital energy from senseless waste and, failing to solve the problem by trial and error, falls into parabiosis and departs. When the parabiosis subsides, the right combinations for the experimental task can occur, as it remembered when it previously tried something and, most importantly, the intensity of aspiration has fallen sharply. Correspondingly, its worried dominant caused by excessive excitement has calmed and disappeared. To explain the behaviour of the monkey, though it was quite sufficient to use neurophysiological and reflex memorization models, which are in no way connected with thinking, Köhler, introduced, without any grounds, a superfluous essence - thinking, which just complicated the understanding of the problem. Moreover, in experiments with monkeys, the experimenter has an initial setting to confirm the hypothesis of the existence of human thinking and speech in monkeys. That certainly affects the quality of experiments. It is difficult to find differences from the owners of ordinary pets who sincerely believe that the dog, gabbling, tells the owner the intricate stories that he understands. Luckily, animals having natural behaviours likely do not suspect that owners anthropomorphize them.

Comparison of human consciousness and animal behaviour forms the fourth dichotomy in the study of consciousness and its dialectical opposites. The consequence of this comparison is anthropomorphism, a humanization of animals and reduction of human consciousness to the behaviour of animals. From this dichotomy, where there was a high level of consciousness attributed to animals, the development of behaviourism was the logical consequence of an attempt to reduce the role of consciousness.

Behaviourism

Pavlov's experiments on animals showed that reactive behaviour is formed on the basis of unconditioned reflexes and that by such influences it is also possible to develop conditioned reflexes that will cause new patterns of behaviour. These results effected the formation of a new direction in the psychology of human and animal behaviour that shaped psychology at the beginning of the 20th century. The philosophical basis of behaviourism was positivism, founded by the French philosopher Auguste Comte, who believed that the only genuine knowledge is knowledge of objectively observed phenomena. Consequently, scientific research has available only observable facts, which completely excludes from the research toolkit subjective methods such as introspection.

The leaders of pragmatism, Dewey and James, influenced the emergence of behaviourism with their philosophical and psychological ideas. Dewey actively developed functional psychology, defined as the doctrine of mental operations as opposed to the structuralist doctrine of mental elements. Operations perform the role of intermediaries between the needs of the organism and the environment where the main purpose of consciousness is accommodation to the new surroundings. The organism acts as a psychophysical whole, and therefore psychology cannot confine itself to the domain of consciousness. It should strive in various directions to the whole variety of connections of the individual with the real world. James Angell wrote in 1910 that the term 'consciousness' would eventually disappear from psychology, as happened with the term 'soul', and that it would be much more useful to just forget about consciousness and instead objectively describe the behaviour of people and animals. All these studies and also the work of Pillsbury, Montague and Meyer prepared a creation of behaviourism whose beginning is associated with John Watson and his famous 'Behaviourist Manifesto' in which Watson outlined the main features of his new psychology. Watson and his followers reduced psychic phenomena to various forms of behaviour, understood as the totality of the body's responses to stimuli from the external environment. Behavioural psychologists focused on the phenomena of behaviour and in terms of behaviour alone, it began to interpret all the phenomena of psychic life.

They interpreted behaviour as a set of reactions to stimuli of the external environment. Such an interpretation in itself opened up broad prospects because, according to their ideas, knowledge of the nature of the stimulus makes it possible to foresee the corresponding reaction. Conversely, by the nature of a reaction, one can judge the stimulus that caused it. Therefore, using the necessary stimulus to encourage some reactions and suppress others, one can achieve the desired behaviour. By announcing consciousness as an insignificant problem, behaviourists could not offer an answer as it arose, which was well described by the famous phrase of Huxley made long before the emergence of behaviourism 'How it is that anything so remarkable as a state of consciousness comes about as a result of irritating nervous tissue, is just as unaccountable as the appearance of the djinn when Aladdin rubbed his lamp in the story.' (Huxley & Youmans, 1869)

Vygotsky bubble

At the beginning of the 20th century the leader of the French school of social psychology, Durkheim (1922) claimed that the individual as part of society, is attached to himself, thereby proclaiming the thesis of the social conditionality of the psyche. However, it was a translation into the language of psychology of the famous philosophical thesis of Karl Marx about the essence of man. 'Feuerbach resolves the essence of religion into the essence of man. However, the essence of man is no abstraction inherent in each single individual. In reality, it is the ensemble of the social relations' (Marx, 1845 (1969)).

A leading psychologist of the Soviet school, Lev Vygotsky, developing the ideas of Marx and Durkheim in the 1920-1930s, put forward the hypothesis of the origin of consciousness due to the internalisation of social relations where internalisation is *'a transition that results in processes external in form, with external material objects, being transformed into processes that take place on the mental plane, on the plane of consciousness; here they undergo a specific transformation – they are generalized, verbalized, condensed, and most important, they become capable of further development which exceeds the boundaries of the possibilities of external activity.'* (Leontyev, 1977 p. 92).

Vygotsky (1978) considered that a person has a special type of mental function, which he called higher psychological function that has a social nature and is completely absent in animals. This function constitutes the highest level of the human psyche, generally referred to as consciousness. Vygotsky noted that throughout the life of an animal the natural environment is acting on the animal, modifying it and forcing it to adapt, whereas a man acts upon nature using tools and thereby alters nature. As a result of his mastery of nature, man has learned to govern his own psyche and thus appeared higher psychological functions such as the ability of a human to bring himself to remember information and to organize his mental activity, expressed in the form of voluntary activity.

Man controlled his behaviour, as well as nature, using artificial tools that Vygotsky called signs. With these signs, primitive man was able to control his behaviour, memory and other mental processes. Signs were substantive, such as a notch on a tree so that the primitive man saw it and knew what to do. A notch on a tree can be meaningfully linked to various types of labour operations; faced with such a sign a person connected it with the need to perform a particular operation. Therefore, these signs served as additional symbols of meaningful work-related operations.

However, in order to perform this labour operation, a person had to think about what he should do. The sign is a means of internal activity aimed at mastering oneself and it is internally oriented. Therefore, signs were the trigger mechanisms of higher psychological functions that acted as psychological tools. Vygotsky believed that in the process of joint labour there was a dialogue between the participants through special signs, which regulated what should be done by participants of the labour process (Vygotsky, 1978).

Vygotsky supposed that the first words of primitive man were probably orders, addressed to other participants of the labour process, which in essence were verbal signs. Primitive man, hearing a certain combination of sounds, performed the corresponding operation. Later, in the course of development, man began to give commands not to members of his group, but to himself. As a result, from an external command function, words were transformed to an internal organizing function. That is, the first function of an ordering person and someone executing these orders was split and the whole process was interpersonal. Later, this turned into a relationship with oneself and so become intrapersonal. The essence of internalisation is this process of the transformation of interpersonal relation into intrapersonal. In internalisation, there is a transformation of external sign-using such as nicks and nodules to the internal such as inner speech.

Vygotsky himself defined the process of internalisation in terms of three transformations. The first one is 'an operation that initially represents an external activity is reconstructed internally. Of particular importance to the development of higher mental processes is the transformation of sign-using activity, the history and characteristics of which are illustrated by the development of practical intelligence, voluntary attention, and memory.' The second is 'an interpersonal process is transformed into an intrapersonal one. Every function in the child's cultural development appears twice'. Initially, between people (interpsychological) and then inside the child (intrapsychological). The third is 'the internalization of cultural forms of behaviour involves the reconstruction of psychological activity on the basis of sign operations.' (Vygotsky, 1978 p. 57-58).

In this way, it may be noted that in Vygotsky's concept, higher psychological functions have mediated the structure and the process of development of the human psyche, characterized by internalisation of human relations and external signs. 'The internalization of socially rooted and historically developed activities is the distinguishing feature of human psychology, the basis of the qualitative leap from animal to human psychology' (Ibid.)

However, despite the creativity of Vygotsky's ideas about signs and internalisation, the genesis of human consciousness remains unclear. Therefore, a pupil of Vygotsky, Leontyev, proposed a hypothesis about the origin of consciousness, according to which the impetus for the emergence of consciousness was the appearance of collective labour process that made possible, and also necessary, the emergence of consciousness. Leontyev (1981) believed that any collective enterprise involves the division of labour and different members of the group begin to perform various operations. In this case, one operation immediately leads to a biologically useful result; others do not give such a result, and serve only as a precondition of the achievements of someone else. Considered without regard to joint activities, such operations are biologically meaningless.

Leontyev gave an example of a primitive collective hunt where a beater taking part in the hunt was stimulated by a need for food. What was his direct role in the hunt? Probably, he aimed to frighten a herd of animals and send them toward other hunters that hid in wait. That is what the result of the activity of this man should be and his activity finishes with that. The other hunters complete the remaining part of the hunt. The outcome of his specific actions does not in itself lead to satisfaction of the beater's need for nutrition. The goals of his actions did not match with the motive of his activity and one can say, for example, that the beater's activity is the hunt but the frightening of game is his action. So, for the first time, under the conditions of collective labour there are such operations that are not directly targetted at the subject of need: the biological motive, but have in mind an intermediate result. Within the framework of these individual activities, the result becomes an independent goal.

Thus, for the subject, the purpose of activity is separated from its motive. In terms of mental reflection, this is accompanied by an experience of a sense of action. For a man to be urged to perform an action that leads only to an intermediate result he must understand the connection of this result with the motive. That is, to discover its raison d'etre (Ibid). However, the discovery of meaning is already the work of consciousness. As Leontyev noted, 'Personal sense, like the sensory fabric of consciousness, does not have its own nonpsychological existence. If in the consciousness of the subject external sensitivity connects meanings with the reality of the objective world, then the personal sense connects them with the reality of his own life in this world, with its motives. Personal sense also creates the partiality of human consciousness' (Leontyev, 1978 p. 135). That is, according to Leontyev, one can assume that primitive people had a personal sense in an unformed consciousness; in other words, one can say that they became a personality not having consciousness.

Moreover, empirical data accumulated by ethology show that different animals have a similar group behaviour during the hunt. For example, a pack of wolves divides into two parts upon discovering the prey or knowing in advance about its location. Some hide in ambush, others become beaters. The ambush takes place in the path of the likely course of the scared victim. The raid pursues the victim with an interception along the way. The method relies on the desire of many animals to flee from the pursuer, not in a straight line, but in a circle. Some wolves chase the prey, others move crosswise, when the victim deviates to the side. Interceptors are usually fewer in number than pursuers. Often, predators chase the victim, moving along parallel courses. In this case, the relay of persecution is taken up by teammates on that flank, in the direction of which the path of the victim's movement has shifted. The coherence of actions in such collective hunting is very great. Switching roles between chasers and interceptors saves the strength of the hunters.

From a logical point of view, a necessary indicator of the conformity of the hypothesis to that fragment of knowledge, on the basis of which it is put forward, is the coherence of the hypothesis with established knowledge. One can see that Leontyev's anthropogenesis theory has logical contradictions within itself and contradicts the numerous data of ethology. Therefore, one cannot accept this paradigm for building a modern theory of human origin. Soviet psychology proposed the consideration of the question of the relationship between the influence on the psyche of innate and social factors, introducing consciousness as the internalization of social relations, where consciousness appeared in the form of an 'ensemble of social relations'. This ended with a series of endless scholastic discussions where the real concept of the origin of consciousness was absent.

Unconscious

Along with consciousness, there are such states that are beyond the boundaries of consciousness and do not reach the proper degree of intensity in order to attract attention. From a physiological point of view, unconscious processes perform a kind of protective function: they unload the brain from the constant tension of consciousness where it is not necessary. A man could not act intelligently if all the elements of his life activity demanded awareness simultaneously.

The unconscious as a phenomenon is often described in literature. For example, Leo Tolstoy gracefully noted this phenomenon in the novel 'Anna Karenina'

> *The sweat in which he was bathed refreshed him; and the sun, burning his back, his head, and his arms bared to the elbow, gave him force and tenacity for his work. More and more frequently the moments of oblivion, of unconsciousness of what he was doing, came back to him; the scythe went of itself. Those were happy moments... The longer Levin mowed, the more frequently he felt the moments of oblivion, when his hands did not wield the scythe, but the scythe seemed to have a self-conscious body, full of life, and carrying on, as it were by enchantment, a regular and systematic work... While he was engaged in this labour there*

The problem of the unconscious interested ancient philosophers. For example, Plato believed in the presence of hidden, unconscious knowledge in the soul, about which a person himself does not even suspect anything at all. He believed that a man would not look for something that he does not yet know, if he had not previously had it unconsciously in his soul. Two thousand years later Leibniz formulated the concept of the unconscious, which was the lowest form of spiritual activity. He believed that all conscious actions arise in an unconscious mind and that, together with the most prominent conscious representations, there are sleeping representations.

The German philosopher Eduard Von Hartmann in his principal work 'The Philosophy of the Unconscious' attempted to combine, in a coherent theory, various ideas about the phenomenon of the unconscious. He treated the concept of the unconscious as a vital force, manifested in instinct and reflexes, as the healing principle of nature, opening up in feelings, character, in morality and leading man to happiness; he believed that the unconscious imperceptibly controls the human mind.

The psychological investigation of the unconscious was initiated by Wundt who likened consciousness to the field of view, in which there is a sphere of the clearest vision, bordered by circles that recede to the periphery with diminishing clarity of sensation and experience, until their complete extinction. He distinguished between the focus of consciousness, where the object is clearly recognized, and the rest of the field of consciousness, where it is either unconsciously perceived or not at all realised. The boundaries between these points are relative and mobile. Wundt believed that perception and consciousness are based on unconscious logical processes and he tried to establish a connection between the laws of logical development of thought with unconscious phenomena, asserting the existence of not only conscious but also unconscious thinking.

At the beginning of the 19th century, Johann Herbart, a German philosopher and psychologist, one of the founders of scientific pedagogy, introduced a dynamic characteristic of the psychic and was one of the first to note that incompatible ideas can come into conflict with each other. In this case, weaker psychic phenomena are forced out of consciousness but continue to influence it. One can say that he was at the origin of the psychoanalysis of Sigmund Freud who transferred the discussion about the unconscious and its place and role in the behaviour of a person from narrow professional psychological exotics to the philistine environment of the Western world. He believed that he brought the modern man to the top of the volcano and made him look into the boiling crater of the unconscious. Freud said that one of the foundations of his conception of the unconscious is Schopenhauer's ideas: 'We have unwittingly steered our course into the harbour of Schopenhauer's philosophy. For him death is the "true result and to that extent the purpose of life", while the sexual instinct is the embodiment of the will to live.' (Freud, 1961, p. 50).

According to Freud, mental activity is like an iceberg, most of which is hidden under water and is controlled by underwater currents. Like an iceberg, the most important part of the mind is the part that a man cannot see. In the mind there are not only conscious but also dark, even wild, elements that are driven underground by reason and social norms and wait only for a moment of weakness and fear to manifest their demonic powers. His doctrine affirms that the human psyche is dominated by the irrational ego constantly under the influence of other structures and it is not independent. In psychic existence, there is continuous confrontation between the conscious and the unconscious, and unconscious sexual desires have a strong influence.

The controlling element of the superego, personifying the authority of parents, represses the lower instincts from consciousness into the sphere of the unconscious, where they accumulate. However, the retention of repressed desires in the unconscious requires the expenditure of psychic energy, which results in the exhaustion of the psyche and the development of neuroses; the neurotic patient suffers from memories, the past in a neurosis is stronger than the present, hence the existence of obsessive traumatic dreams, phobias and experiences.

Freud paid much attention to the analysis of dreams that he considered to be 'the royal road to the unconscious' and understood as the repression of primitive antisocial drives while awake and their symbolic satisfaction in a dream. Freud believed that obsessive dreams are neurotic states of the normal person, performing protective functions. He showed the presence of similar scenes in free associations as in dreams with mythical scenarios of the ancient epic.

Jung, Freud's disciple, expanded the Freudian concept of the energy of sexual drives to psychic energy not related to the somatic sphere, but they differed in the understanding of psychoanalysis because of Jung's refusal to accept an exclusively sexual interpretation of the motivation of human actions. He believed that it was impossible to explain the reasons for the loss of contact with reality in schizophrenia and other mental illnesses solely by sexual repression. In addition, Jung expanded the structure of the unconscious, highlighting within it the personal unconscious, which affects the emotional sphere of the individual and the collective unconscious that is inherited patterns of behaviour that includes the superpersonal experience of humankind. The collective unconscious is key to Jung's theories of the mind and consists of archetypes (universal primitive images) and ideas. The archetype is a primary structural element of human psyche.

Neumann (2015) noted that any attempt to outline the archetypal stages should begin with a fundamental delineation of personal and transpersonal mental factors. Personal factors are those that are characteristic of one individual and are not shared with anyone else, regardless of whether they are conscious or unconscious. Transpersonal factors, on the contrary, are collective, suprapersonal, and should be perceived not as external social conditions, but as internal structural elements. The transpersonal is a factor that does not depend on the personal factor since the personal one is a later product of evolution. In his opinion, neurosis is a consequence of reproducing, in the human mind, archaic experiences and images incompatible with modern reality. Thus, a person is born with a ready personal sketch, and the environment only brings out already incorporated capabilities and features. Each person at birth inherits the structure of the psyche, developed over hundreds of thousands of years and it is this structure that causes the individual to live through his experience in a certain way.

Jung employed Haeckel's law of recapitulation, in which the biological development of the individual replicates that of the species; he wrote 'The supposition that there may also be in psychology a correspondence between ontogenesis and phylogenesis therefore seems justified. If this is so, it would mean that infantile thinking and dream-thinking are simply a recapitulation of earlier evolutionary stages' (Jung 1967, para. 26). In this context, Neumann (2015) noted that the gradual evolution of consciousness equally applies to humanity as a whole, as well as to the individual. Therefore, ontogenetic development can be considered as a modified repetition of phylogenetic. Not all stages of the development of consciousness are present always, everywhere and in every mythology. The theory of evolution does not state that all stages of development of all species of animals are repeated in human evolution. What can be said is that these stages of development represent an organized sequence and direct all mental development. These archetypal stages are predetermined by the unconscious and are found in mythology. Only by considering the collective stratification of human development along with the individual stratification of the development of the conscious can we come to an understanding of mental development in general and individual development in particular.

The boundary between the theories of Freud and Jung is a line between ontogenetic and phylogenetic approaches. Freud's complexes have a basis in the history of personality development, whereas Jung's archetypes have a basis in the paleohistory of the human race, which can be expressed through mythology. Therefore, it can be said that archetypes, as constituent elements of the collective ideal, have a foundation in the psyche of individuals and it is possible to find a rudiment of the paleo-psychic in the psyche of modern man by studying his deviant behaviour.

The unconscious, as we have noted, is a mental process that does not enter the sphere of consciousness and which affects human behaviour without the subject realizing the determinants of this behaviour. By analysing the behaviour, the researcher is informed about the unconscious, however, the basic nature of the unconscious is not revealed by the researcher. The unconscious can be characterized by insensitivity to logical contradictions and the lack of a linear irreversible sequence between the past, the present and the future.

At first glance, the unconscious appears as the dialectical opposite of consciousness, as antonyms. However, from the point of view of dialectical analysis, it is difficult to accept this conclusion. Since the various schools of psychoanalysis examine behaviour and motives, we are not talking about consciousness and the unconscious, but rather about unconscious and conscious behaviour. Psychoanalysis has never set itself the task of researching the emergence of consciousness. Therefore, there is no significant antinomy of the concepts of 'conscious' and 'unconscious'.

Conclusion

There are a large number of definitions of consciousness in various philosophical and psychological schools; many of them are built on the idea of a dichotomy when the concept is divided into two incompatible concepts. For example, in philosophical definitions, consciousness as something material is opposed to the ideal. In psychological definitions, consciousness is opposed to the unconscious; in communication theory, the verbalized consciousness is often contrasted with the non-verbal consciousness. Medical research considers a dichotomy between clear consciousness and altered states of consciousness. Often used to describe the consciousness are qualitative and quantitative characteristics such as the volume of consciousness, the stream of consciousness and the multiplicity of different consciousnesses. Since all these definitions somewhat contradict one another, there cannot exist any phenomenon corresponding to all the encountered definitions of consciousness. A historical view shows that there are several major problems associated with consciousness that are considered as mental-physical, mind-body and mind-brain dichotomies, to which one can cite most points of view on the nature of consciousness.

For example, at the philosophical level of analysis, Priest (1991) distinguishes several major and relatively independent approaches to understanding consciousness. There is dualism (Plato, Descartes, Popper and Eccles), logical behaviourism (Hempel and Wittgenstein), idealism (Berkeley and Hegel), materialism (Davidson and Honderich), functionalism (Putnam) and double aspect theory (Spinoza, Russell and Strawson). Finally, there is the phenomenological view (Brentano and Husserl).

In psychology, consciousness is a major subject in that it is independent of how the attitude to the problem of consciousness in a particular school is declared. The development of the science can be represented in the form of successive attempts to find the origins of consciousness and to determine consciousness through its dialectical opposition. Therefore, the subject of research in different paradigms can be imagined as a set of dichotomies of consciousness and its dialectical antithesis.

The first dichotomy was a psychophysical problem (the dichotomy of consciousness and matter) the problem of the relationship of the psychic with the material that Descartes posed. In connection with the development of physiology and anatomy in the nineteenth century, the aforementioned psychophysical problem was transformed into a psychophysiological one, which considers the relationship of the psyche to the processes that occur in the nerve substrate. The second dichotomy became a psychophysiological problem of the consciousness of man and the anatomical vessel in which it is carried.

The doctrine of reflexes combined physiology and behavioural problems. Researchers represented consciousness in various forms in the early studies of reflexes, for example, in the form of an unfinished reflex, a complex reflex or as a sum of reflexes. In the end, the researchers came to the conclusion that reflex and conscious actions are not only dissimilar, but in opposition.

Consciousness does not arise where more and more complex reflexes are being developed, but where there is a rejection of the instinctive-reflex, animal mode of action. Reflex and consciousness do not transform into each other; it is impossible to understand the beginning of consciousness by studying the reflex. Comparison of the reflex and consciousness forms the third dichotomy in the study of consciousness.

Comparison of human consciousness and animal behaviour forms the fourth dichotomy in the study of consciousness and its dialectical opposites. The consequence of this is anthropomorphism, the humanization of animals and reduction of human consciousness to the behaviour of animals. From this dichotomy, where there was a high level of consciousness attributed to animals, the development of behaviourism was the logical consequence as an attempt to reduce the role of consciousness. The most recent was the dichotomy of consciousness and unconsciousness, which was introduced into the scientific discourse by Hartmann, Herbart and Freud.

Chapter 7

Genesis of consciousness

In the previous chapter, the traditional research paradigm of the origin of consciousness was considered which is based on step-by-step logic of the formation of human society. Supposedly, this complex form of existence arises directly from a less complex form through gradual development, without interruption, without negation of negation, and without a period of chaos through the direct continuation of the development of biological forms of lower levels. At the heart of the new approach, which received a rigorous scientific exposition recently and synthesises the theoretical achievements of different schools, there are studies in different fields of knowledge that unite the negation of the purely gradual logic of the formation of human society and the recognition that a complex form of being arises from a similarly complex form by inversion. For these modi operandi, the name 'inversion approach' is proposed. A great influence on the development of the inversion approach to the problem of the emergence of consciousness was introduced by Kretschmer in 1921, in his theory of psychogenesis and Lévy-Bruhl (1923), who was one of the first to note the opposition of pralogical and logical thinking. Later, Porshnev's works in the field of sociogenesis, the Chertok's studying of hypnosis and the Bekhtereva's theory of a stable pathological condition of the brain made an excessive contribution to the evolution of the inversion theory.

Primitive thinking

Lévy-Bruhl

French sociologist Emile Durkheim introduced the important scientific concept of collective representations. The totality of beliefs and feelings, in his opinion, are the same for members of the same society, forming a system that has its own life. According to Durkheim, people have two consciousnesses. One contains only states peculiar to each person and it is characterised as individuality. The states covered by the second are common to the whole social group. Collective consciousness or collective representations are not granted to a person from his immediate experience, but imposed on him by a social environment.

Starting from the ideas of Durkheim, Lévy-Bruhl, a French philosopher and anthropologist, developed the idea that the human consciousness is socialized and contains a lot of traditionally perceived collective representations, with different forms of thinking corresponding to different societies. Lévy-Bruhl (1923) declared that primitive populations, living in the modern era, think in a totally different way from westerners.

At that time, the approach of the British anthropologists, Tylor and Frazer, was dominant. According to this, primitives think the way modern people do, just less scrupulously. In their approach, the disparity between primitive and modern thinking is one only of degree, whereas for Lévy-Bruhl the difference was of kind. He states: 'Let us abandon the attempt to refer to their mental activity to the inferior variety of our own' (Lévy-Bruhl, 1985, p. 76).

He states that primitive peoples are not objectively inferior to their modern counterparts and they have faith that all occurrences are part of an impersonal sacred that everyone participates in. 'Primitive man, therefore, lives and acts in an environment of beings and objects, all of which, in addition to the observable properties that we recognize them to possess, are endued with mystic attributes' (Lévy-Bruhl 1985, p. 65). He gave an example of the people from the Brazilian Bororo tribe who declare themselves to be red parakeets. This does not indicate that after their death that they become parakeets, nor that parakeets are metamorphosed. 'It is not a name they give themselves, nor a relationship that they claim. What they desire to express by it is actual identity' (Lévy-Bruhl 1985, p. 77). The Bororo trust that humans and parakeets are indistinguishable and at the same time distinct.

Lévy-Bruhl came to the conclusion that the logical laws governing collective representations of primitive people do not at all resemble the laws of thinking of modern people. For primitive thinking, the external reality itself is mystical. It does not seek to explain the phenomena of the surrounding reality, as these phenomena are perceived separately, and in combination with the magical properties of objects. Collective representations include strong emotions and the performance of religious rites, sharply excites the nervous system, causes fear, horror and passionate desires (Ibid.).

In the thinking of primitive people, their own special laws prevail. The place of the logical laws of identity, and contradictions, is taken by the phenomenon that Lévy-Bruhl called the law of participation. An object can be both itself and at the same time something else. It can be here and at the same time in another place. A person feels himself mystically united with his totem and his forest soul. Calling such a type of thinking pralogic, he stipulates that this is not the same as the illogical. In pralogic thinking there is no propensity for analysis, when the synthesis is carried out directly, without identifying the causes of the event, therefore the cause of the event may be any factor in time or space. Lévy-Bruhl gives an example: two anklets were found in a crocodile stomach by a European hunter. Later, the aboriginals identified that these items belonged to two women who have disappeared recently. However, instead of concluding as modern people that the crocodile had on its own eaten the two women, the aborigines deduced that some sorcerer 'had summoned the crocodile, and had bidden it catch the two women and bring them to him', as though crocodiles would never eat people unless bidden to do so and the anklets were its reward (Lévy-Bruhl, 1923, p. 52).

The perception of the world under the dominance of collective ideas is differently oriented than the perception of a modern person who strives for the objectivity of cognition. Lévy-Bruhl noted that the product of perception in a primitive man instantly envelops a certain complex state of consciousness, in which collective representations dominate. Therefore, they do not distinguish between dreaming and reality, the person and his image, the person and his name, or the person and his shadow.

However, in order to avoid misunderstandings, Lévy-Bruhl (1923) repeated that all these features of primitive thinking are inherent only in collective representations when logic folds and pralogical thinking enters into force with its mysticism and the law of participation. In the sphere of his personal practical experience, primitive man argue and act similarly to modern man. Moreover, they have hyper-developed memory and remember many more trivialities in comparison with modern men that use memory, in general, to preserve the results of logical thinking.

Jung, who was already discussed, adjusted the Lévy-Bruhl's approach. He offered the psychological interpretation of primitive thinking that hypothesised that primitive peoples think as they perform as they live in unconsciousness. Jung proposed that primitive thinking is the original psychological state of all human existence. Primitive peoples do not distinguish what is inside from what is outside and they project themselves exclusively on the physical world that is managed by divine forces. 'For primitive man the psychic and the objective coalesce in the external world... Primitive man is unpsychological. Psychic happenings take place outside him in an objective way' (Jung, 1970, para. 128).

Dialectic of insanity and sanity

Kraepelin and Schneider

Schizophrenia is a severe progressive endogenous mental illness that begins on the basis of a hereditary predisposition, characterized by disharmony and loss of unity of mental functions, leading to peculiar personality changes in the form of autism, reduced energy potential, emotional impoverishment and increased introversion. Schizophrenia is one of the most compound and mystifying diseases to affect humanity. It is a relatively common mental illness with an approximate prevalence of around 1% in the general population and half of long-term psychiatric patients in mental hospitals (Hirsch & Weinberger, 2003). Approximately 90% of cases of schizophrenia occur at the age of 15-54 years. The incidence does not depend on nationality or race and there is no human community free from schizophrenia (Ibid.). According to Frith and Johnstone (2003), each year approximately 8% of patients with schizophrenia commit acts of violence. This is more than in people without psychiatric disorders, in which this figure is approximately 2%. Other studies show that from 5% to 10% of murders in Western countries are carried out by people with schizophrenic disorders (Simpson 2004). The study of Palmer *et al* (2005) estimates that 4.9% of schizophrenics will commit suicide, usually in the early stages of the illness, a considerably higher risk than in the general population.

Even though the term schizophrenia is comparatively new, the idea of psychosis is very old. Schizophrenia-like mental disorders in which thought and emotions are so impaired that contact is lost with external reality have been recognized since at least the era of Ancient Greece. Greek tragedy portrayed heroes who are tortured by psychosis and who were often driven to performing horrendous actions while insane. In the Euripides' play, the Bacchae Agave murders her son King Pentheus, persecuted by the delusional belief that he is a lion. Later, William Shakespeare in his plays such as Hamlet, Othello, King Lear and Macbeth portrayed persons who experience schizophrenia. Schizophrenia as an independent disease was identified in connection with the creation of the first nosological classification of psychoses and the transition of psychiatry from the symptomatic to the nosological period of development.

Kraepelin (1919) offered a general concept of psychosis divided into two main groups. The first group of patients (manic-depression) who were psychotic had a short course, usually with a full remission whereas the second group of patients had a chronic course and typically progressed to a deteriorated condition (dementia praecox). Thereby he formulated the theoretical basis that generated the contemporary concept of schizophrenia. In 1908, Eugen Bleuler described schizophrenia as an independent disease, different from dementia and introduced the term into psychiatry. Its main feature is not dementia, but a violation of the unity of the psyche, including a violation of associative thinking. Bleuler singled out how the diagnostic criteria of four As: decreased affect, autism, lack of associations and ambivalence that was considered by Bleuler (1950) as the main signs of schizophrenia.

Later in the 1920s, Kurt Schneider identified fundamental symptoms of schizophrenia; he suggested that loss of the sense of personal autonomy and an incapability to find the boundaries between self and not self are its critical components (Schneider, 1959). His psychotic symptoms have entered into the modern diagnostic classifications of mental disorders DSM-IV as positive symptoms of schizophrenia and included delusional perception, auditory hallucinations, thought disorder and passivity experiences. He considered that the identification of schizophrenia should, as a rule, be based on the first-rank symptoms. That included: 'thoughts experienced as spoken aloud or echoed; thought withdrawal, thought insertion and thought broadcasting; voices heard commenting on the patient's thoughts or actions and voices discussing the person in the third person; feelings, impulses and volitional acts experienced as under the control of some external force or agency; somatic passivity and delusional perception.' Schneider attempted to discover hidden symptoms of schizophrenia that are secondary causative. However, as noted by Pull (2002), he displayed only the existence of dissociation of psyche that may support the diagnosis and there are many various manifestations of schizophrenia.

Schneider noted that patients with schizophrenia often have unusual perceptions. It can be illusions when actual objects are perceived to be distorted and hallucinations when there is a perception without an object such as auditory or visual hallucinations. The patients attach special importance to combined hallucinations, since they are perceived as most plausible. For example, a patient can see a cat, hear its meow and feel the touch of its soft fur. Around half of patients with schizophrenia have experience of auditory hallucinations when a person hears voices talking to the patient or among themselves. Often, the voice is not associated with anyone known by the patient and it can be identified as male, female or few different voices. These voices are as a rule understandable and experienced as coming from the external. Typical for schizophrenia are voices that replicate the patient's thoughts and talk to the patient in the third person.

The endogenous origin of consciousness

Ernst Kretschmer

Ernst Kretschmer was Kraepelin's student and one of the most significant psychiatrists in the first half of the 20th century. Kretschmer was primarily interested in the relationship between the etiology of mental disorders and the occurrence of symptoms of disease in terms of physical and psychological characteristics. He presented an unconventional nosological concept: the idea of the constitutional body type and established the well-known trifold typology of the leptosome, athletic, and pyknic habitus.

The unpredictability of manifestations of schizophrenia and the absence of a direct connection with actual reality led Kretschmer to the idea that the origins of this phenomenon should be sought in the past, that is that there is a phylogenetic determination of schizophrenia. He noted that abstract thinking considered from the evolutionary point of view could not be wholly derived from visual images of the external world (Kretschmer, 1922). One can say that his idea of the endogenous origin of consciousness formed a new paradigm of the origin of consciousness, as most researchers sought external determinants of the origin of consciousness.

Recall that Lévy-Bruhl proposed a law of participation as the basic principle of primitive thinking. However, many psychiatrists note that it is remarkably similar to the syndrome of schizophrenia splitting of the psyche. According to the law of participation, a primitive man easily admits that one and the same creature can reside in several places simultaneously; that in one person different entities can be settled; that one essence can live in different people and that different entities can settle in different parts of the same body. However, exactly the same descriptions can often be found in the clinical descriptions of schizophrenia.

Therefore, it is no accident that European colonists often perceived primitive people as crazy. However, one can take into account the members of the Bororo tribe (which we mentioned earlier) assuring that they directly and currently are red parrots or a Papuan that kills a neighbour because after sunset at his hut is a howling dog, and this is a sign of death, so he had to go and help a neighbour with the transition to the valley of the dead.

In a modern society, people with this kind of thinking become psychiatric patients. However, it is possible, as Ten (2011) supposes that they are normal people which are not trapped in their era, carriers of pralogical thinking that has recapitulated in them as an atavism. However, in the wild, they could have survived more successfully than any resident of modern London or Paris, which is proved by numerous examples when patients of psychiatric clinics were in the forest due to various circumstances, such as war, were better adapted than ordinary people.

Kretschmer paid attention to the analysis of dreams and he noted that during sleep, 'I' and the external world break up and turn into each other, that people cease to distinguish them from each other. In the dream, at some point the 'I' appears, at the next moment and even simultaneously with the first one, there may be a 'not-I'. He spoke of the separation of personality during a dream, the personality disintegrates or splits (Kretschmer 1922).

We noted earlier that the most reliable symptom of schizophrenia is the dissociation of the psyche. However, as noted Kretschmer, the same is observed in the dreams of perfectly normal people. Creative, ingenious people, especially artists and poets, so tirelessly conduct an analogy between the way their works appear and dream that this can be regarded as a firmly established fact. Their works appear much more easily in a state of mental half-darkness, with a lowered consciousness blunted by external attention, as in a state of distraction with a hypnotic-like concentration on one point, or a passive experience, often of a sensual character, accompanied by oblivion of space and time and renunciation of logic and will (Ibid.).

In this phase of artistic creativity existing close to a dream state, early phylogenetic tendencies to rhythm and stylization are awakened. In the same way, human emotion is built in a state of hypnosis which is passive, like dreams, and the experiencing person experiences a sense of being a spectator where the images are projected with a distribution along a spectrum between the 'I' and the external world. Often, one can readily find an affective agglutination of otherwise incompatible images, different from in everyday life, from primitive man's psychology and the psychology of dreams (Ibid.).

Analysing the psychopathic manifestations of twilight disorder of consciousness and schizophrenia, Kretschmer made a somewhat paradoxical conclusion that there is no major mechanism related to images or affects, as we described them in primitive man, which we could not find in schizophrenics. As Ten (2011) further assumes, schizophrenia is a consequence of the atavistic activation of ancient neurosystems.

Despite that, Kretschmer proposed the idea of the endogenous origin of consciousness and together with Lévy-Bruhl noted the opposition of primitive thinking and the thinking of a cultural man, he did not offer a clear answer to the question what could have been at the origins of the consciousness of modern man. Perhaps, this was because he remained within the paradigm of the gradual development of the psyche from animals to man.

Reversing inverted

Boris Porshnev

Russian researcher Boris Porshnev (1974) tried to answer the question of what could have been at the origins of the consciousness of modern man and, since the 1930s, developed an original theory of anthropogenesis whose essence can be expressed as the following: consciousness could not have appeared simply as a result of the gradual development of the animal psyche but rather in its formation it has passed through some turning points, undergoing an inversion, meaning the formation through 'its other'.

Porshnev proposed an original concept of anthropogenesis. In his opinion, anthropogenesis is not a process of gradual humanization of ape-like ancestors, but a sharp turn over an abyss during which something fundamentally different from monkeys and people appeared in nature, and then disappeared. He believed that the labour of man is fundamentally different from the work of Pithecanthropus and Neanderthals, whose work was similar to the work of the beaver and not Homo sapiens.

Porshnev argued that the ancestor of man could not be a hunter, as this contradicts the data of zoology and the only niche that he could occupy was scavenging. A large number of archaeological finds of tools for cutting animal carcasses supports this theory. However, the making of these tools is analogous to weaving a web, building nests and beehives, blocking the river with dams of other animals that adapt to the specifics of their ecological niche and it does not speak about the presence of consciousness. However, the ability of primitive people to conquer the niche of carnivorousness in the fight against such scavengers as the American vultures or hyenas raises doubts. Moreover, necrophagy requires a specialized digestive system, such as scavenger vultures have, that on the one hand has a high acidity of the digestive tract and on the other hand, has evolved an extraordinary tolerance to bacterial toxins in decaying meat. Roggenbuck *at al.* (2014) show that there is a strong adoption of vultures and their bacteria to their nutrition, with which they have a specialised host–microbial symbiosis. Nothing of this sort has been seen in primitive man. The idea of Porshnev that a primitive man could eat brains from under the skull or in tubular bones is interesting enough but this could occur at a rather late stage of anthropogenesis. One can say that Ten (2011) corrects him considering that, in the beginning, primitive people consumed molluscs and only later began to devour the brains that they extracted with the help of stone tools.

Porshnev, answering the question of why at some stage of the evolution of biological systems suddenly appears a social system, formulated an interesting philosophical paradox: The social cannot be reduced to the biological, yet social characteristics cannot be deduced from anything but the biological. He offered a solution to this antinomy that based on the idea of inversion. The last may be expressed in a few words as follows: 'a certain quality (A / B) is transformed in the course of its opposite (B / A): nothing is new here, but everything is new.'

At the beginning of human history, everything was reversed from what it is now. 'The course of history represents the gradual reversal of the original state. However, this original state was in its turn preceded by another inversion: a reversal of the animal nature into the one from which humans have started their history.' Consequently, the story completely falls under the formula of Feuerbach 'reversing inverted' (Porshnev, 1974, p. 13–17).

Porshnev noted that, at the dawn of history, people's mental characteristics were not only dissimilar to that of modern humans but represented their total opposite and he sharply criticized attempts to cast a gulf between man and animal both from the human side in comparisons with animals, but much more so from the animal side as anthropomorphisms. This approach does not so much raise the problem of the transition from animal to human as it tries to show that there is no special problem; it does not resolve the task but removes it entirely. In the frame of this approach, there is the explanation of psychogenesis: bit by bit, the psyche has become more complicated; bit by bit, the brain grew and gradually herd relations developed into human social relations. This is self-deception. It is necessary to look with open eyes at the fact that the transition from the zoological level to the human level has not yet been explained (Ibid.).

He believed that thinking at the beginning of anthropogenesis makes the body more helpless than animals and asked a logical question of how, if we exclude all mysticism, can we explain this useless feature? After all, natural selection does not preserve harmful features. Consequently, he suggested that thinking could develop only as a result of artificial selection. Arhanthropus (*Homo erectus*) built unique symbiotic relationships with numerous predators, herbivores and birds. A rigid instinct did not allow them to kill anyone. Therefore, the use of animal biomass in food was limited to the use of animals that had been killed by predators. However, because of the reduction in the food supply, they had to become predators and start killing. Two opposing instincts combine: 'kill' and 'do not kill'. One of the solutions to this paradox was the killing of representatives of its own kind, as the instinct did not forbid it. That is, the use of a part of its population as a self-reproducing fodder source. This phenomenon of adelphophagy occurs in zoology, for example, intrauterine cannibalism of sharks, in which the strongest embryo actually consumes its smaller womb-brothers. The phenomenon sometimes reaches considerable size in some species, although still never becomes the main source of nutrition.

As a result, a division of the species of Paleoanthropus (possibly Neanderthals) began to occur into two species. From the previous species, a new one broke off relatively quickly and violently, becoming an ecological opposite. If Paleoanthropus did not kill anyone other than their own species, then the new species of neoanthropes represented an inversion: as they turned into hunters, but they did not kill Paleoanthropus. Consequently, Paleoanthropus became cannibals. They consumed the most helpless representatives of their species and gradually they began to breed representatives of the 'eaten branch'. Subsequently, of this 'selection', the most helpless step by step became even more helpless, but with ever larger brains. Later, the neoanthropes took advantage of their large brain and thus emerged intellectual advantages that generated social relations and modern man.

He also proposed an original approach to the development of higher mental functions. He believed that provoking imitation could inhibit another animal from any action without the aid of positive or negative reinforcement and at a distance connect the inhibitory dominant and imitation. Porshnev considered the yawn and smile as stimuli that block an action by making a targeted animal copy its antagonist by means of spontaneous imitation. He supposed that the preliminary function of this behaviour was interdiction (a distant neurosignalling effect of one individual on another) for the suppression of the partner's excessive reactivity by inducing inhibitory dominants.

An example of defensive interdiction in the herd may be a situation when the ringleader that is trying to give a command is suddenly interrupted as members of the herd distract him by, for example, scratching his head or yawning or falling asleep, irresistibly provoking the law of imitation. Discovering the interdiction as a way of signalling the impact on the tribesman, the human ancestor began to spread this practice in relation to other animals. That allowed them to occupy a unique ecological niche and build a symbiotic relationship unprecedented in the animal world. Paleoanthropus, having thoroughly polished his skill in the field of interdiction on other animals, began to apply this powerful tool to other Paleoanthropus. Thus, the circle traversed by the interdiction closed: formed within large clusters of Paleoanthropus and adapted for use solely in relation to other animals, the interdiction returned to the internal relations of the Paleoanthropus.

Despite the creativity of the idea of interdiction, Porshnev could not find from where the interdiction comes. Why do normal animals begin to behave inadequately, forgetting reflexes, abandoning instincts? Evolution with its adaptive mechanisms adapts and further specializes all inadequate behaviour or rejects them. In its mechanisms, there is no place, where inadequate reflexes could be accumulated and so one day donate the whole collection to a single species. Porshnev, taking only the external determination of consciousness and not accepting the idea of an endogenous factor, could not offer a convincing mechanism for the emergence of consciousness.

At the beginning of anthropogenesis, in the Porshnev concept is asserted, the second signal system was to the first signal system in full functional and biological antagonism and was formed on the foundation of the negation of the first. It raises fundamental questions: what can cancel the machine-like automatisms of the first signalling system? Is there a mechanism of the brain at the level of the first signal system (in the reflex mechanism), anything to which the expression 'contrariwise' would fit? In fact, Porshnev believed that in the reflex mechanism and the mechanisms of inhibition, one can find the sources of human consciousness. Developing the ideas of Vvedensky and Ukhtomsky on conjugate inhibitions, he proposed the existence of a conjugate inhibition in some central region of the brain coupled with a dominant centre for explanation of the nature of inadequate reflexes. Thus implying, as the primary condition for the emergence of the consciousness of the existence of two centres (excitation and inhibition) in one brain.

He noted that mental illness in people is inevitable, as according to the laws of genetics, individual features of ancestral species play in a certain small percentage of human beings. Those people who strongly reproduce some mental features of the ancestral form, often become patients of psychiatric clinics and their symptom complex is a monument to the life of beings who were the forerunners of modern people. This idea is close to the Kretschmer concept that the human psyche passed in its phylogenetic development through the stage of dissociated consciousness, a stage of insanity. Porshnev pointed to some ancient secret, enclosed in a split personality; and the sources of mental illnesses are contained in the phylogeny of consciousness, namely, in its original duality, dissociation. Here we again return to schizophrenia where the key symptom as repeatedly noted by many authors is the dissociation of consciousness.

Stable pathological condition of the brain

Bekhtereva

If one considers mental problems more widely as pathopsychology, that is predominantly revealed by anomalous cognitive, perceptual, and intellectual functioning in the course of mental illnesses, it can be noted how altered self-awareness in the course of the reflective activity of the diseased brain generates a new activity for man: the activity of self-perception. This activity, due to connection with the constancy of unusual feelings and their special significance for a person, becomes meaningful and takes the lead in the hierarchy of other activities. Patients cannot think of anything except their own unusual conditions and their causes (Zeigarnik, 1986). This frenzied activity of self-knowledge, manifested in the pathogenesis of a mentally ill person, as Ten (2011) notes, certainly has phylogenetic roots. After the phylogenetic inversion of the psyche, when the social principle prevails and demands its tribute in the form of observing conventions of social life, unusual feelings activity of self-perception fall into the category of pathology along with many other kinds of primary mental activity.

In this context, one can also note Bekhtereva's theory of a stable pathological condition (SPC) of the brain as an adaptation base for many chronic nervous system ailments that has opened principally new opportunities in the treatment of these diseases. Bekhtereva suggested that with a chronic brain disease, a kind of new homeostasis develops, ensuring optimal adaptation of the diseased organism to the environment (Bekhtereva, 1978).

It is important to keep in mind the main factors of the SPC: a general reorganization of the state and interaction of brain and organism systems; the further maintenance of this reorganization by practically the same reactions of the organism that previously held the homeostasis of health. Maintenance of the reorganization is based on the newly formed long-term memory matrix. Since the state is always a huge complex of processes, therefore, memory is needed not for one event, but an interconnected and complex memory matrix. Disease as a process, the course of the disease should be distinguished from a stable pathological condition that is a persistent result of the disease, but both these aspects of the disease should be treated in unity, despite their qualitative difference.

Bekhtereva believed that the adaptation of the organism to internal conditions that changed as a result of the disease occurs not only by replenishing the damaged brain systems but also by forming a new homeostasis, a new stable and self-sustaining pathological condition. Even when the factor causing the problems disappears, the body often cannot get out of this state by itself.

A SPC is one of the most important factors of adaptation of the organism to the environment and from a biological point of view, it is positive. However, in treatment it is a complicating factor since the body begins to consider this pathological condition as correct. Therefore, the transition from the SPC to the normal state is usually accompanied by a phase of destabilization and temporary deterioration. This is the reason for the stability of the SPC; the body struggles against deterioration and the patient himself does not understand this and subconsciously makes efforts to not be taken out of a state of stable pathological balance (Ibid.).

The SPC of the psyche originates in the archaic form of adaptation of pre-sapiens to the conditions of functioning of their psyche and development towards Homo sapiens. In modern people, it appears as a relapse. Arising as a result of various factors, it tries to take the place of normal consciousness and regain its lost functions. Perhaps during the formation of human consciousness in phylogeny, a stable pathological condition was an adaptation to the conditions of brain functioning, which develops as a synthesis of the dual structure. Therefore, having deep phylogenetic roots, it is extremely stable. Unlike man, in animals, this principle cannot be observed because they have nothing to recapitulate, for them any mental abnormality e.g., rabies is a disease leading ultimately to death.

A stable pathological condition is something restored, a recapitulation of a post-animal and pre-human. This is something with which it is possible to live and if not for modern ideas about the normal, this anomaly would be the highest form of the psyche, as it once may have been. One can say that this is not a disease, but an evolutionary malfunction, a violation of the correct course of the brain. This is the slipping of consciousness into a more stable and specialized, but simultaneously more primitive, state. The adaptation and stability of the human body to specific conditions of the external environment can increase. However, this short-term gain led an individual or a group of individuals into an evolutionary impasse where they could be completely happy but lost the prospect of development.

However, it is important to note that if these abnormalities are registered with a small frequency in a society then they are perceived as a disease. However, when there are a large percentage of such deviations, they cease to be evaluated as a disorder and the pathological condition itself ceases to frustrate patients. As soon as such people become the overwhelming majority, the mentally abnormal becomes the previous normal state.

To understand this idea, one can consider an old parable that, in different versions, occurs both in the East and the West. One day an angel came to a prince in a dream and said that this night the water on the Earth will become poisoned, and everyone who drinks it will go mad. The angel advised the collection of a large quantity of clean water and to drink only this until the environmental water clears up. Waking up, the prince jumped out of the house and tried to warn everyone he could about the trouble to come. But people did not believe him. They laughed and drank water. And they went crazy. He tried at least to save someone. But everyone turned away, not wanting to listen to him. With horror, he looked at the world around him crumbling. But those who drank the poisoned water did not notice this, because everyone around them was just like them. Only some crazy prince ran around the streets and shouted about the poisoned water and about how the world was until today. The prince did as the angel told him, and drank pure, clean water. All continued to think him crazy, and he waited for the water to clear. Time passed and the prince became tired of living alone in an existence of misunderstanding and mockery, so he drank water from a local well. Since then, there has been a rumour about a wonderful well, the water from which healed the crazy prince.

It is interesting that there are cultures where people refuse to consider the thinking of the majority as the norm, but the thinking of the minority as a psychiatric pathology. For example, in the Russian Chukotka or in the Alaskan tribes of the Eskimo, the shaman is a schizophrenic in terms of modern medicine. From the point of view of the Eskimo, the shaman is the chosen one of the spirits, the mediator who had the ability to see a different reality and travel there. Certainly, the shaman departs from the social norms of his society, but he has a high status, and no one considers the isolation of shamans despite their abnormality.

Hypnosis and the psyche of pre-sapiens

The dissociation of the human psyche was also studied using various types of hypnosis when hypnotized people observed phenomena characteristic of an altered state of consciousness: positive and negative hallucinations, age regression, post-hypnotic suggestions and many other effects. The main characteristics distinguishing the hypnotic state from the ordinary state of consciousness is a loss of sense of control over normal reactions (Hilgard, 1991) and the presence of insensitivity to contradictions of logic (Orne, 1979). This is close to pralogic thinking.

Cognitive psychology explains hypnosis in terms of a mismatch between cognitive subsystems and a control system that carries out conscious behaviour management (Hilgard, 1991). According to his approach, many cognitive subsystems can be identified in the psyche, each of which has a certain degree of consistency, sustainability and autonomy. These subsystems are hierarchically ordered and constantly interact with each other, but in hypnosis, they can become partially isolated from each other. To manage these subsystems, there is an executive ego. Hypnosis changes the hierarchical order of subsystems, which causes violation of motor control, distortion of perception and memory, and the emergence of positive and negative hallucinations.

In the framework of analytical psychology, the main triggering mechanism of hypnosis is the transference. The meaning of the transference is as follows: during the immersion in hypnosis, the hypnotized person temporarily refuses from the innate mechanisms of self-defence and vigilance, giving his identity to the hands of another (Chertok, 1982).

Initially, it was thought, that the transference is the cause of hypnosis but the cause of the transference was suggestion. That is, the process occurs according to the scheme: suggestion - transference of the hypnotized person to the hypnotist - the hypnotic state, trance (altered state of mind). However, the study of hypnosis in the 20th century has showed that transference is not transfer of personality to a hypnotist; the transfer involves a third party. Hypnosis includes a well-known experience, a tangle of phantasmatic and symbolic representations. Therefore, one can pose a logical question: whether the relations developing in hypnosis are much more archaic relations, strictly dual, which existed before human activity. We are referring to the initial biological state, rooted in the animal world, a state in which communication is carried out on a purely affective level, not yet related with a figurative content (Chertok, 1982).

Hypnosis is the immersion in the initial state of the psyche, which Chertok calls a dual within the psyche of a person undergoing hypnosis and he notes that it is not a singular primitive consciousness is being reconstructed, but a split form. Being under hypnosis, a person moves to more primitive forms of communication, to affective functions that correspond to the functions of the most archaic structures of the nervous system. That is, he speaks about the existence within a personality of an innate potential ability to establish relationships within itself, which forms an unfilled matrix into which all subsequent systems of relations will fit.

The idea that hypnosis is the restoration of a state that was actually previously experienced, included in the phylogenetic heritage that reproduces the relation to the primitive father was suggested by Freud in the Group Psychology and the Analysis of the Ego (1921), however Chertok (1982) revealed deeper phylogenetic layers of personality. He called hypnosis the fourth state of the body along with the states of wakefulness, sleep and the activity of dreaming. This later received instrumental confirmation in Talbot's (2008) study who noted that, unlike other states of altered consciousness, hypnosis is not associated with any deviations in the electroencephalogram. From the point of view of physiology, the state of the psyche in hypnosis resembles the usual state of wakefulness. Does this mean that ordinary state of wakefulness is a kind of hypnosis?

Ten (2011) believes that two persons are encoded on the ancient matrix: 'I' and 'someone' and the hypnotist in this system of relations is the third party, acting as mediator and causing the 'spirit of the lamp'. He raises some questions: in what struggle did the consciousness of man form, if the traces of this struggle are still recorded on the 'matrix' and whence did autism and inner speech appear in the hypnotic state? From where did the paleo psychic of man's ancestors have such a complex dual structure?

He believes that in the depths of our consciousness, in its most ancient layers, transference occurs constantly, continuously and intensively, which removes (or perhaps cleans) the internal dialogue. Without the existence of this transference, people would wander like sleepwalkers and a hypnotic dream in reality would be a normal condition. From this, logically, Ten (2011) concludes that there are at least two different variations of 'I' in man. At the same time, one of them constantly dominates, continuously presenting the second 'I' with a demand of transference. The transference is a tribute to the dominant and the expression of readiness to obey at any moment. Only thanks to the internal transference, people retain the integrity of their perception and thinking.

An indispensable condition of internal hypnosis is the presence of more than one 'I', or inner person, in a man. Trying to find the answer to the genesis of these two, he suggests that in antiquity there was a period when the struggle between the two persons and emotional inner dialogue was the usual state of the psyche. At the dawn of consciousness formation, the normal state of hominids was unfamiliar to people's wakeful consciousness with its integrity of perception, but rather a hypnotic somnambulism with its own duality, when order and performance were separated in time and when the order sounded like that of a hypnotist.

This state differs from the state of modern man who does not hear such internal commands and do not notice the time that is shared by the order and the performance. This is the norm. However, in special pathological cases, there are people who hear the commands, performed somnambulantly and repeated loudly. This is an external manifestation of intense internal dialogue. As a rule, the actions of these people are slow; they are as if trapped in a state of hypnosis. When the disease progresses, the second 'I' gets out of control and can begin to impersonate anyone.

As an example, such a picture can be observed with schizophrenia as was already mentioned and with dissociative personality disorders, in which a man's personality is divided and there is an impression that there are several different personalities in the body of one person, each with a completely different nature. Moreover, it does not depend on cultural differences. For example, Suryani & Jensen (1993) note that the phenomenon of brightly manifested trans-states in the community of Bali has the same phenomenological nature as multiple personalities in the West. The dissociation of consciousness is visibly reflected in post-hypnotic spontaneous amnesia. Not accidentally, amnesia and the problem of multiple personalities have long been at the centre of thinking about hypnosis. This phenomenon manifests itself in the testimony of people who have undergone hypnosis, who often say they have experienced a state of depersonalization or a split personality, as if all this was happening to someone else. Statements in various forms such as 'It seemed to me that it was not me, but someone else who acted' are often found in the descriptions of experiments with hypnosis (Chertok, 1982, p.115).

The inversion model

Different approaches to analyse to the problem of the origin of consciousness may be attempted. Most authors agree that consciousness is the highest form of the development of psyche, which differs from psyche of animals. In animals, psyche is functioned by the type: external stimulus – reaction where the reaction is predetermined by instincts or based on external experience or trained reflexes so that this reaction can be always expected. In humans, external stimuli enter into the consciousness, are handled by a system of questions like 'What do I need? What response do I want to give to someone who sends me a signal of aggression?' This is called self-reflexion and this reaction of man can be non-standard, unusual and unique, which normal animals do not have. If my already mentioned cat Liza, in response to the usual stimuli, responded with original reactions then I might suspect that consciousness awoke in her.

Analysis of the definitions of consciousness shows that most researchers give definitions in the genetic discourse; they describe the development from simple forms such as chemotaxis (the movement of bacteria in response to a chemical stimulus) to increasingly complex ones. It is like giving a definition of wheat, describing how it grows, from which grain and under what conditions. They avoid giving definitions in the essentialist discourse.

In addition, most researchers note that consciousness is an active reflection of reality that processes within itself external stimuli. Researchers successfully study the psyche of organisms from the simplest manifestations, for example chemotaxis, to the reflexes of higher animals.

Nonetheless, they typically stop short of tackling the problem of the emergence of consciousness that actually separates animals and man.

Some researchers find an answer in theology or esotericism that proceeds from the idea that the human consciousness is a cast of the Supreme consciousness, or His function, or His reflection. But while introducing the concept of a higher power and not explaining its genesis, they leave the logical field of science. Most concepts of scientific, rather than theological, origin of consciousness believe that consciousness has an external source: a reflection of the surrounding world. A fundamentally different approach is possible when the cause of the origin of consciousness is communication between the different structures of one brain, so one can talk about the endogenous origin of consciousness.

The evolutionary approach considers that nature produces more and more complex forms of existence, and then it breaks down and produces something fundamentally new, and this new thing develops again into something even more complex. Thus, the brain appears: initially, the nerve fibres, then the lump, next the lump is divided in two, which means the separation of functions and the possibility of competition. The highest form of development of the divided brain is recorded in dolphins, in which exists the most tortuous terminal cerebrum. It is dominated by longitudinal furrows that do not extend from one hemisphere to the other, so that the hemispheres are not interactive to the same extent as in men or apes. The hemispheres of the dolphin's brain are much less dependent on each other, and represent almost two autonomous brains, only one of the hemispheres rest and not the entire brain of the dolphin. This explains the inherent state of wakefulness. These two brains are forced to work independently of each other, cooperate, coordinate and compete with each other. Perhaps it is when dolphins are involved in the simultaneous work of both hemispheres of the brain that there is a violation of harmony (one can say dolphin schizophrenia) that leads to the fact that dolphins beach themselves on the shore. The cause of the violation could be underwater earthquakes or technogenic activity such as ultrasound of sea-going vessels.

A human brain is a further stage in the evolution of the brain, when two independent brains are brought together into one organ and actively communicate and compete within it. The interactivity of the hemispheres, ensuring the unity and speed of the human brain's work, has been established by the development of the callosal commissure, which is plexus of nerve fibres in the brain, connecting the hemispheres, through which they exchange information.

This evolutionary acquisition, through which the activity of human consciousness differs from the dissociated work of the dolphin's brain, was formed as a result of the return of the *Archaeocetes* to land. Within the framework of the inversion theory, which is introduced in the previous chapters of this book, it has been disclosed that the ancestors of man were most likely a neotenic form of Acrodelphinidae. So, the shallowing of the sea and the return of the *Archaeocetes* to the land led to the fact that both hemispheres, practically autonomous brains, were able to sleep and stay awake at the same time. As a result, they began to conflict with each other, tearing the psyche apart, which led to both the emergence of insanity and the origin of consciousness. We noted that when the two dolphin brains are turned on at the same time a collapse occurs and the dolphins are thrown onto the land. Drawing parallels with dolphins, one can say that when two brains of a primitive human were turned on simultaneously, this being was thrown onto 'the coast of the mind', but not immediately, but through an intermediate instance - schizophrenia.

The solution of the problem of the onset of consciousness should not be confined to the possibility of a direct transition from the instinctive-reflex mechanisms of the animal psyche to human consciousness or, using the Pavlov terminology, cannot move little by little from the first signal system to the second signal system according to nondialectic logic. The basic idea of the study of human evolution is the consideration of the phenomenon of human consciousness, not as a result of the slow, progressive development of the animal psyche but as a transformation, an inversion, a decisive turning-point of the psyche from the animal from which man originated. Pavlov was one of the first to note the need for an inversion approach to the problem of the emergence of consciousness; he wrote that the consciousness of man is not the result of a slow progressive development of the psyche of animals but its basis lies in the transformation, the decisive turning point, and the reverse course of evolution. Pavlov's conclusion can be considered as the methodological basis of the inversion theory of psychogenesis.

When the phenomenon of hypnosis was discussed in this chapter, it was noted that hypnosis is an immersion into the original state of the psyche, which can be considered dual-natured. Therefore, one can assume that the human consciousness is split and has no unity of consciousness; the unity of perception and thinking peculiar to people in their normal state is guaranteed only by the obedience of one part of the split structure of consciousness to another and by the rejection of its own initiative in favour of the other. In other words, at the dawn of the formation of consciousness, the normal state of hominids was not wakefulness with its wholeness of perception, but a hypnotic dream. It was an internal dialogue, where a command and its execution belonged to different minds living in one brain.

At the stage of formation of the human brain, there was no consent, so our ancestors were schizophrenic and autistic. This often caused internal group conflicts. Researchers find remains of anatomically modern people of one tribe who died as a result of bloody group conflicts. Perhaps the cause of these conflicts was the schizophrenic states of the members of the group that have been tearing the psyche apart, which had led to both the emergence of insanity and the emergence of consciousness.

Numerous populations of pre-sapiens died, unable to withstand this mental tension, only a small population managed to survive passing such an evolutionary bottleneck. Perhaps there were only two of them: Adam and Eve. Passionate love could become that imperative vector that allowed the split of the psyche to be overcome because man as a creature has a unique hypersexuality, characterized by the strongest, all-consuming, emotional attachment to a sexual partner. When society was formed, it began to dominate the psyche of individuals, cleansing it of schizophrenia.

The core idea of the inversion theory is that man has received the ability of active reflection of external stimuli, because he has an internal mirror that never, until a man in consciousness, stops self-reflecting. Thus, the ability to reflect external incentives is an acquired characteristic of man and not the cause of the emergence of consciousness, as many philosophical schools believe.

It can be suggested that the endogenous theory of the origin of consciousness fits much more organically into the scientific picture of the world and, accordingly, it is possible to give a completely essentialist scientific definition of consciousness, defining it as the highest form of the psyche, characterized by its ability of self-reflexion. The criterion for the presence or absence of consciousness is self-awareness. The structure of self-consciousness can be characterized by how a person realizes his differences and distinguishes himself from the outside world as he realizes himself as a subject capable of changing the surrounding reality and he begins to realize his own mental processes and properties. A person who thinks he has consciousness, in fact, has it, because this very thought is the criterial manifestation of consciousness. No animal, even the most intelligent, can turn the psyche upon itself. This would mean that it is capable of analysing its behaviour and its own psyche and engage in self-criticism.

Ten (2011) notes that the criterion of scientificity in the question of the origin of consciousness must be a predicate of inversion, as a dialectical leap. The challenge is to reach out to the causes of the breakthrough in development, that break in gradualism when the development of the genus Homo sapiens went against the natural determinants when natural determinations were removed from the interior by evolution itself. Since evolution has not a formal logic but rather a dialectic logic, where nature sublates itself in order to develop but does not stagnate in the equilibrium of highly specialized species.

The fundamental novelty of the inversion theory lies in the fact that it links morphogenesis and psychogenesis into a single entity. The rationale for morphogenesis is the foundation of the theory, but the main element is the inversion theory of psychogenesis, which states that consciousness appeared as a result of an inversion dialectical leap (discontinuity of the continuity of quantitative changes) in evolution, and not as a result of the step by step complication of monkey reflexes since the emergence of consciousness on the basis of reflexes is impossible, as proved by neuroscience during the 20th century. A possible reason for the inversion was the malfunction of the harmonious animal psyche of pre-sapiens into a schizophrenic state. The anatomical basis for this malfunction was the situation when there were two autonomous brains in one organism, which man inherited from his ancestors.

From the evolution point of view an inversion was a step backwards, because adaptability sharply decreased, pre-sapiens were on the verge of self-destruction. It was a step comparable to charging the AK-47, when the shutter is retracted before the shot which leads to the emergence of consciousness. On the basis of a split consciousness, thanks to society, consciousness was formed. Society here is the main causal factor of the occurrence of consciousness, but this is not the same as the instinctual sociality of so-called social animals. Human sociality has a vector of ongoing improvement because people are able to format society, whereas in animals instinctual society is predetermined by nature and unchanging.

Conclusion

Levy-Bruhl established that the logical laws governing collective representations of primitive people do not at all resemble the laws of thinking of modern people. In their primitive thinking, their own special laws prevail; there is no propensity for analysis, when the synthesis is carried out directly, without identifying the causes of the event, therefore the cause of the event may be any factor in time or in space and they do not distinguish between dreaming and reality, or the person and his image.

Schneider described the similar pattern when he identified loss of the sense of personal autonomy and an incapability to find the boundaries between self and not-self as critical components of schizophrenia. Kretschmer offered that there is a phylogenetic determination of schizophrenia and the abstract thinking, considered from the evolutionary point of view, could not be wholly derived from the visual images of the external world. In addition, he made the conclusion that there is no major mechanism related to images or affects, as we described them in primitive man, which we could not find in schizophrenics. Ten assumes, in fact, that schizophrenia is a consequence of the atavistic activation of ancient neurosystems.

Porshnev proposed that anthropogenesis is not a process of gradual humanization of ape-like ancestors, but a sharp turn over an abyss, during which something appeared in nature, and then disappeared. He confirmed Kretschmer's idea that the human psyche passed in its phylogenetic development through a stage of dissociated consciousness, the key symptom of schizophrenia, a stage of insanity.

Chertok supposes that in hypnosis, a split form of consciousness is being reconstructed. Being under hypnosis, a person moves to more primitive forms of communication, to affective functions that correspond to the functions of the most archaic structures of the nervous system.

Ten believes that transference occurs constantly and intensively in the depths of our consciousness, which removes the internal dialogue. At the same time, there are two different variations of 'I' in man; one of them constantly dominates, continuously presenting the second 'I' with a demand of transference. At the dawn of the formation of consciousness, the normal state of hominids was not wakefulness with its wholeness of perception, but a hypnotic dream. It was an internal dialogue, where a command and execution belonged to different minds living in one brain.

The cause of the origin of consciousness is communication between the different structures of one brain, so one can talk about the endogenous origin of consciousness. Man has received the ability of active reflection of external stimuli, because he has an internal mirror that never, whilst a man is in consciousness, stops self-reflecting. Thus, the ability to reflect external incentives is an acquired characteristic of man and not the cause of the emergence of consciousness. The basic idea of the study of human evolution is the consideration of the phenomenon of human consciousness, not as a result of the slow progressive development of the animal psyche but as a transformation, an inversion, a decisive turning-point of the psyche from the animal from which man originated. The scientific theory of the origin of consciousness must be based on two fundamental postulates. The first is the primordial dissociation of the psyche of pre-sapiens; as a result, the formation of Homo sapiens looks like a dialectical synthesis, as a process of development through contradiction, a process of becoming the unity of opposites, which consciousness is in essence. The second postulate is the idea of Ten (2005) about the endogenous source of consciousness, which follows from both the practice of studying the phenomenon of split consciousness and the idea that each hemisphere can have its own separate thinking (Sperry, 1961) and from the idea of phylogenetic development of the two hemispheres with a certain degree of independence (Crow, 1997).

Chapter 8

Applied aspects of inversion theory

Different phenomena such as visionary experience can sometimes be combined with the phenomenon of synaesthesia, which is an interesting psychological phenomenon that can produce new ideas for understanding the origin of consciousness.

Altered state of consciousness

The phenomenon of distinct trance states as well as the phenomenon of multiple personalities in the West is usually accompanied by visionary activity that as empirical descriptions are widely present in literature and art. There are many descriptions, which tell of such visions, and of people who have a supernormal ability to perceive a kind of hidden reality. Visions are inextricably linked with the development of mankind, which ascribes specific cultural and historical meanings to the visionary. Virtually all religions have evidence of a visionary experience. For example, the experiences of Buddha, recorded in Jataka or the texts of the Old Testament (the books of Daniel and Isaiah).

Dronfield determines visions as subjective visual phenomena that are formed inside the neural network of the visual pathway and have their cause at different points between the retinas and the cortex. The phenomenon includes entoptic, entophthalmic, phosphene and hallucinatory visual experiences and practically everybody has had some experience of this phenomenon, such as spots before the eyes, seeing stars or flares when rubbing or strongly closing the eyes (Dronfield, 1993). The fuller spectrum of the phenomena of the visionary experience can only be seen in altered states of consciousness, such as hallucinogenic intoxication, trance, stress, sensory deprivation or stimulation with oscillating light, or in some psychiatric illness such as epilepsy and schizophrenia (Ibid.).

The first detailed phenomenological description of visionaries in the form of philosophical reflection was written by the English writer Aldous Huxley in his essay 'The Door of Perception' (1954). He discovered a similarity of the characteristic features of different in origin states. He compares the perception of the world, changed by the reception of mescaline, the impression of the works of art of different schools such as Blake and van Gogh and a description of the visions of medieval mystics. Thanks to his interpretation, the visionary takes on the features of a universal phenomenon with stable characteristics. He used hallucinogens that are able to make radical changes in the perception of space, time, and the self; also, it is possible that they depersonalise a man. Colours can be perceived as extremely bright, and objects of the environment lose their meaning and begin to live their own life. The person that is under the effect of a hallucinogen is separated from society and is no longer subject to the values of that society.

In 1988, Lewis-Williams & Dowson said that at all times people had a similar nervous system, and the constancy of certain encountered visual symbols can be explained, rather not by the cultural symbolic tradition, but the antiquity of the human nervous system and its generation of entoptic phenomena. Typical examples of these entoptic phenomena are 'luminous, pulsating, expanding or contracting, blending and changing geometric forms. They include zigzags, dots, grids, meandering lines and U-shapes.' (Lewis-Williams 1995, p. 6)

Analysing the specifics of these geometric patterns, Lewis-Williams suggests that these patterns left by trancing shamans may be some artefacts of the nervous system and become the key to understanding the state of consciousness which the shamans were experiencing. In this context, Dowson notes that rock art is somewhat even more extraordinary than signs of the past; it offers us a window into human cognitive evolution (Dowson, 1994). Different researchers propose that in the Paleolithic, prehistoric artists painted these pictures because they thought it would enable a successful hunt. However, it was found that there was an insignificant connection between the food and the paintings of the primitive people (Laming, 1959). Therefore, images on the walls of cave of all these animals' heads, horns and rushing herds are not carefully drawn from nature, but the product of entoptic visions.

There are many hunter-gatherer cultures including North-American and African tribes, which create rock art and have used hallucinogens such as plants of the Henbane family and psychoactive mushrooms, such as the Amanita muscaria for a long period of time. Possibly, Paleolithic cultures could have used similar plants as hallucinogens and it is also possible that hallucinogens were the reason why people started creating rock art because entoptic phenomena have been recognised in their paintings. These appear particularly bright when under the encouragement of hallucinogens. Lewis-Williams **(2002)** shows the resemblances between the entoptic phenomena defined in laboratories and recognised in San rock art and in Paleolithic rock arts. Also, from the 1950s, attention was given to resemblances between abstract configurations, influenced by hallucinogenic substances or electrical stimulation, and patterns in African art.

Based on the research of Lewis-Williams, Laming, and Leroi-Gourhan (1982) one can say that the origin of art is an altered state of consciousness. This is confirmed by fantastic figures, and unthinkable combinations of symbols that could not be copies of something seen in the normal state of consciousness. Furthermore, the location of some works of primitive art is in such caves, where it is impossible for outsiders to view. This means that the rock drawing is not intended for the general public but for the artist himself where two of his inner 'I' could consider the art. That suggests the autism of artists and supports the hypothesis of their altered state of conscious.

Many authors note that one reason for the generation of images could be the state that is described by psychiatrists as schizophrenia. However, it is not possible to determine precisely what was at the source of the altered state of consciousness of primitive artists: the phylogenetic dissociation of consciousness or the use of plant hallucinogens or the combined effect of these two factors. From the inversion theory of consciousness it can be assumed that in the psychogenesis of man, in which human ancestors existed in a situation of phylogenetic schizophrenia, psychotropic substances could be a catalyst for the process of changing consciousness.

Synaesthesia

Visionary experience can sometimes be combined with the phenomenon of synaesthesia, which is an interesting psychological phenomenon that can give new ideas for understanding the origin of consciousness. It is well known that the human nervous system is endowed with different channels of perception: sight, touch, smell, hearing, taste, a sense of balance and position in space and so on. Each of these has an established level of autonomy, and the perceptual channel is activated through specific external stimuli such as sound, light and pressure. Sense organs do not always work in isolation and can interact with each other; the work of one sensory organ can stimulate or suppress the work of another sensory organ. There are also interactions in which the senses work together, causing a new kind of sensitivity, which in psychology is called synaesthesia. This can be described as 'the involuntary physical experience of a cross-modal association. That is, the stimulation of one sensory modality reliably causes a perception in one or more different senses' (Cytowic 2002).

For example, it is widely known that people perceive high sounds as bright and low as dark. The same applies to smell; some smells are evaluated as weightless and light, while others are heavy or dark. In art, there are facts of coloured hearing, which is experienced by many people and is particularly evident in some musicians, such as the composers Scriabin, Franz Liszt, Jan Sibelius and Duke Ellington. This synaesthesia was most likely possessed by physicist Richard Feynman, philosopher Ludwig Wittgenstein and writer Vladimir Nabokov. Synaesthesia is met more frequently than people recognize. According to legend, St Augustine discovered that when a reader reads silently, he hears the sounds of the words without any audio input. This is a case of hearing synaesthesia.

Cytowic (1993) estimates that 40 people in one million are synesthetes and women synesthetes prevail with a ratio of 3:1 (in the US). He found that synesthetes have normal standard neurological exams and are normal in the conventional sense. They have excellent memories and they cite their parallel sensations as the cause of memories. People with synaesthesia do not shine in mathematics, but they have very good memory and strive for order in literally everything. They are strikingly good at remembering the spatial location of objects such as the location of tools, furniture arrangements and floor plans, books on shelves, or text in a book. Possibly related to this observation is a predisposition to favour order, preciseness, and balance (Ibid.). Most synesthetes are left-handed or ambidextrous and they are also inclined to creative professions and art. Synaesthesia is often inherited: 40 per cent of synesthetes have a close relative, who also experiences the blurring of boundaries between sensations. The synesthetic experience is irrational and often unconscious; however, special exercises allow reflection upon the process. There are data on the possible manifestation of synaesthesia in humans with altered states of consciousness in some forms of schizophrenia, with meditation, hypnosis, hypnagogic states, physical exertion and as a reaction to certain medications.

Russian neuropsychologist Alexander Luria (1968) described the case of Solomon Shereshevsky who was a newspaper reporter in the 1920s, and his editor noticed that he never took notes when he interviewed news sources. The puzzled chief sent him to Luria who subjected the reporter to a range of tests and was amazed by the results. Later, Luria wrote:

> *'He read or listened attentively and then repeated the material exactly as it had been presented. I increased the number of elements in each series, giving him as many as thirty, fifty or even seventy words or numbers, but this, too, presented no problem for him... when the experiment was over, he would ask that we pause while he went over the material in his mind to see if he had retained it. Thereupon, without another moment's pause, he would reproduce the series that had been read to him.'*

Shereshevsky could memorize a sequence of items and narrate it either forward or in reverse; he did not even need to know what he was memorizing, for example, he memorized a stanza from Dante's The Divine Comedy in Italian while he did not speak Italian, and was able to recite it from memory with faultless inflection 15 years later. He also easily recalled the list of words and circumstances associated with it 15 years later. For the duration of the experiment he closed his eyes and said: "Yes, yes ... you gave this list to me once, when we were in your apartment. You were sitting at the table, I'm in a rocking chair ... you were in a grey suit and looked at me like now... now, I see how you said."

Luria diagnosed in Shereshevsky an extremely strong version of synaesthesia. When he heard music, the notes would activate flashes of colour, and touching something induced him to feel taste. In addition, he had an astonishingly bright imagination, one able of conjuring up extraordinary images even from seemingly unremarkable numbers. For example, when thinking about numbers he reported, "for me 2, 4, 6 and 5 are not only numbers. They have form. 1 is a sharp number, independent of its graphical representation, it's something finished, hard... 5 – complete finishing in the form of a conic, tower, fundamental." Shereshevsky was a polymodal synesthete and could hear "the bell ringing… a small round object rolled before my eyes… my fingers sensed something rough like a rope… I experienced a taste of salt water… and something white." (Luria 1968).

Shereshevsky could visualize every memorable word, but this created difficulties for him in reading, because each word generated a vivid image that interfered with the understanding of what was read. His perception was very specific, words expressing abstract concepts, for example eternity, presented special difficulties for him, since they are difficult to compare with the visual image. He could not fully control the memory, it worked unpredictably. When he tried to memorize a set of words written on a board, other sets of words written on the same board before came to mind and all of them got mixed up in his memory. Therefore, he had to constantly make efforts to suppress certain images, making room for others and this forced him to invent special techniques to forget.

Unexpectedly, Shereshevsky had a poor remembrance for faces and voices heard on the telephone. When he tried to remember someone's face, his mind was filled with images of this person, imprinted with memory at different moments of life in minute detail, in all sorts of angles, with different lighting and in all variety of expressions. Therefore, it was very rare for him to recognize a person. He complained: "The faces are so changeable; I am constantly confused by the endless shades of their expressions".

Shereshevsky's memory obeyed the laws of perception and attention rather than the laws of memory, he could not reproduce a word if he had not seen it well. The recall depended on the illumination and size of the image, on its location, on whether the image was obscured by a stain that arose from an extraneous voice. Luria illustrates this by extracts from protocols. 'Reproducing a long row of words S. missed the word pencil. In another row he missed the word egg.' Shereshevsky explained his errors: "I placed the pencil near the fence – you know the fence outside – and the pencil merged with the fence and I walked by it… The same thing happened with the word egg. I had put it up against a white wall and it blended into the background. How could I possibly spot a white egg up against a white wall?" (Luria 1968, p. 36-37). Reading was a difficult and tedious job as he struggled to get through the visual images, which, in addition to his will, grew around every word. Understanding of complex and abstract logical and grammatical structures was much more difficult for him than for people with an ordinary memory. Luria noted that his synesthetic images control his thinking and his mnemonic ability turned into a type of secondary pathology of the psyche.

Shereshevsky had enormous difficulties dealing with synonyms or metaphors caused by the images that crowded in on him. For example, words like child, youngster and infant bring many dissimilar things to mind. Additionally, a homonym, that is one of a group of similar words that have different meanings such as 'wear' as in 'to wear away' or 'to wear a coat', created difficulties because the image would always be identical even though the meanings were different.

Whenever Shereshevsky communicated with someone, a synesthetic visual perception began to control his thinking, which created difficulties with the formation of concepts and generalization. The associations were easily recreated when he recalled the initial situation which was recorded in his memory so he did not need a logical organization (Ibid.).

Luria noted signs of a split personality in Shereshevsky who said: "I needed to go to school... I saw myself here, while 'he' had to go to school. I'm angry with 'him' - why 'he' is going to take so long?" And then he remembered, "I'm eight years old. I stay in the house - that is, I see how 'he' is standing by the window in my old room. He does not go anywhere." According to Luria, this split between the 'I', which orders, and 'he', this carries out orders, accompanied Shereshevsky throughout his life. In the notes, Shereshevsky said, "Imagine the situation. I'm sitting in your apartment, absorbed in my own thoughts. You, as a hospitable host, ask: 'How do you like these cigarettes?' – 'To be honest, so-so.' That is, I would never have answered, but 'he' can. It's not tactful, but I cannot explain it to him. 'I' understands this, but 'he' does not. As soon as I digress, 'he' immediately says what he should not say."

One can say that there is a dialogue between the two personalities of Shereshevsky. The first 'I' lives in the ordinary world and with limited memory. The second 'I' exists in an altered world in which there is unlimited memory and perhaps pralogic thinking. In the case of Shereshevsky, the second 'I' was part of his ordinary consciousness. Luria, being a professional doctor, describes the history of Shereshevsky case very carefully referring to the object of research anonymously. Most modern psychiatrists could easily diagnose this as schizophrenia.

We noted that Shereshevsky was recalling information after ten years in the same form and with the same detail as on the day it was received. This may indicate that there was no process of transformation and generalization in his memory, that is, logical thinking did not participate in memorization. He certainly did not use any logical resource of remembrance at all and he often did not take note of the sense of many words.

When we described the psychology of the primitive man on the basis of the studies of Lévy-Bruhl, we noted that the product of perception in primitive man instantly envelops a certain complex state of consciousness in which collective representations dominate. Therefore, they do not distinguish between dreaming and reality, the person and his image, the person and his name, the person and his shadow. Similar descriptions we find with Shereshevsky when he describes the existence of two distinct varieties of 'I'. He can be himself and at the same time another person or he can be here and at the same time in another place.

He does not seek the objectivity of knowledge, memorized information is perceived without any analysis, which is analogous to the perception of the world in the dominance of collective representations. It can be said that to a certain extent that in his pralogical thinking there is no inclination for analysis and synthesis is carried out directly. Although, in the sphere of his personal practical experience, Shereshevsky argued and acted similarly to his contemporaries, Lévy-Bruhl noted that the carriers of pralogic thinking have hyper-development of memory such as Shereshevsky and they remember many more trivialities in comparison with ordinary men that use memory, in general, to preserve the results of logical thinking.

Within the framework of the inversion theory of the origin of consciousness, it can be assumed that in primitive Homo the process of memorization was analogous to the memorization of Shereshevsky who used the mechanism of synaesthesia. It would be more accurate to say that the long-term dissociation of the Shereshevsky consciousness was similar to pralogical thinking of primitive people and some forms of schizophrenia. As a result, this created the possibility of not fully integrating the functions of logical analysis of memorized information and this ultimately ensured his extraordinary mnemonic abilities.

Analysis as an operation involves putting information into the Procrustean bed of a more general context, which involves cutting everything down, that does not enter into formed the general concept. This general concept is the final result of synthesis in the process of logical thinking. When the synthesis is carried out directly, without analysis, the memory uses the mechanisms of synaesthesia, no information is cut off and it is firmly imprinted in the brain not as a concept but as a sensually multidimensional image. This image is actualized whenever a person perceives a similar sound, colour or smell. For example, the logical consciousness in which the concept of an apple tree is formed will not necessarily react to the smell of an apple by immediate restoration of the definition, of what is an apple tree. However, the pralogic consciousness, with any hint such as a slight smell, actualizes the image of the apple tree. Therefore, Shereshevsky had difficulty reading books: every word attracted an image, distracting from the meaning of the text.

There are several concepts for explaining the phenomenon of synaesthesia. The first one believes that the cause of synaesthesia lies in the thin axons of nerve cells that are able to exchange electrical impulses with neighbouring axons, which, with a certain signal, cause additional sensations. According to the second theory, synesthetes have specific neural articulations preserved from infancy. That allows infants to distinguish sounds, tastes, colours. That is, every child up to 6-7 months is a synesthete. The third theory suggests that cross-activation between different zones of the cerebral cortex in which the centres of sensory perception are concentrated in a person causes synaesthesia.

It can be noted that these theories reduce individual synesthetic experiences, in which consciousness associates different colours to certain music or letters, to sufficiently simple neurophysiological processes such as incomplete myelination or local network disruption of neurons. That is, for the analysis of such a complex psychic phenomenon as synaesthesia, simple interpretation schemes are used that lie in the sphere of physiology and psychophysiological reductionism.

Cytowic (1993) supposes that synaesthesia is not a higher cortical function in the conventional meaning and one can propose that 'the level of this unknown link occupied a low to intermediate level of the neuraxis, rather than a higher level involving more mental mediation' (Cytowic & Wood, 1982a; 1982b). He found that cortical metabolism plummets during synaesthesia; hemispheric flows are low and inhomogeneous and drop an additional 18%, on average, in the left hemisphere. It is impossible to get such a decline in a normal person with a drug. A physical or mental task, or any activation procedure such as a drug, carbon dioxide or oxygen inhalation, typically increases blood flow by five to ten percent. As a result, he concludes that 'synaesthesia is a cross-association phenomenon which takes place in the limbic system of the left hemisphere' (Cytowic 1994, p.163) and generally decrease cortical activation. It depends only on the left-brain hemisphere and is accompanied by large metabolic shifts away from the neocortex that result in relatively enhanced limbic expression.

The inversion theory allows us to give a more coherent explanation for the decrease in metabolism in synesthetes. It is well known that the human brain consumes a huge amount of energy relative to its mass: approximately 25% of the total energy of the body is used by the brain. At the same time, energy consumption of neurons is comparable with the use of other somatic cells and the main consumers of energy in the brain are their commissures (bundle of nerve fibres). These populations of neurons are analogous to a computer bus that transfers data between the functional units of a computer. The largest brain commissure is the callosal commissure that connects the left and right cerebral hemispheres and consisting of around 250 million axonal projections. In synaesthesia, this mechanism of information exchange, the formation of a common image, and then the general concept are broken. The hemispheres work in a relatively isolated way, like two autonomous brains, and there will be multiple personality disorder as in Shereshevsky case, or one hemisphere in general is inactive and only the hemisphere which thinks in images works. As a result, the non-operation of the commissure reduces the energy expenditure of the brain in synesthetes. Of course, this conclusion confirms the validity of Cytowic that synaesthesia is not a higher cortical function, on the contrary, it is a lower evolutionary level, and this phenomenon can be explained only by recapitulation. In the case of very young infants, this is a normal recapitulation, like infant autism; on the other hand, in the case of adults it is a disease like autism or schizophrenia.

Ramachandran & Hubbard (2001) note that synaesthesia is an innate sensory phenomenon caused by a mutation of certain genes that causes unusual activity of various areas of the brain. A closer relationship between zones responsible for different concepts and activities creates a greater predisposition to imaginative thinking and creativity. Based on statistics that synaesthesia is much more common among artists, poets and representatives of other creative professions, they believe that the synaesthesia genes are the basis of creativity and contain a spectrum of qualities necessary for a person as a species. These genes were preserved in the process of evolution because they empower some people with the ability to connect different ideas (Ibid.). Unfortunately, the genes responsible for synaesthesia are not identified and the hypotheses about genetic mutations can be supported only by the deficit of satisfactory theoretical models when the explanation of the phenomenon of the psyche, and the characteristics of the human individual, are attempted by genetic reductionism.

However, discourse about the relationship of creativity and synaesthesia should be carried out carefully as this phenomenon, as we saw in the example of Shereshevsky, reduces the level of logical thinking which encourages affected people to avoid intellectual activities. Instead, the system of social selection encourages synesthetes to engage in visual arts, theatre and dance, where the logical component is insignificant. Subsequently, this creates a statistical anomaly of a high proportion of synesthetes in these creative professions.

Marks (2011) talks more cautiously about the role of synaesthesia in human evolution and notes that synaesthesia grants a minor Darwinian advantage, if it is associated in a causal way with cognitive processes that make available an additional benefit to information processing. He supposes that synaesthesia is associated with better memory and maybe with creativity.

Cytowic (1993), having analysed an evolutionary development of synaesthesia, supposes that synaesthesia is a neural process that occurs all the time in every human. Highly processed and evaluated information about the world is received from sensory receptors and converges in the limbic system, particularly the hippocampus, for multisensory evaluation. However, this multisensory awareness has been lost from the minds of most people, so Cytowic named synaesthesia a metaphorically cognitive fossil because, perhaps, synaesthesia arose among animals or pre-humans in their prelinguistic phase. Later, he explains his thesis 'Synaesthesia is animalistic, not because it is somehow primitive but because the sensory percepts are closer to the essence of what it is to perceive meaning than are semantic abstractions.' (Cytowic, 2002, p. 10). However, multisensory knowledge has been lost by most people, and now appears as now appears as a certain 'cognitive atavism' (Ibid.). Therefore, one can assume that primitive people had cross-sensory perception that existed simultaneously with split thinking. That is, with a little simplification, one can say that most of them were synesthetic schizophrenics. Later in the course of anthropogenesis, there was a loss of these two features (split thinking and synaesthesia) of the human psyche.

Cross-sensory perception exists in the animal world as an essential characteristic. Evolutionary synaesthesia arises by way of physiological interaction of sensations when a weak stimulus causes an excitation process within the brain structures that easily transfers to other brain structures. As a consequence of the spread of excitation, the sensitivity of other sensory organs can be increased. Simultaneous receipt of information about the environment along several communication channels provides the formation of a complex polymodal image of the perceived stimulus, which increases the survival rate of the species in the animal world.

Marks (1989) notes that synaesthesia is not a phenomenal manifestation restricted to humans only, it can be experienced by animals but it is important to keep in mind that a researcher should consider synaesthesia more or less in sensory terms rather than in cognitive terms. In comparison with animals, humans have specific forms of synaesthesia due to learning and language. Many researchers show a well-defined role of language and semantic networks in the development of several kinds of synaesthesia, especially flavour reactions to words that were presented by either the acoustical or optical modality. Possibly, there are different types of synaesthesia (lower and higher synaesthesia) and animals and man share only the lower types such as the perception of pain in colour and sound in colour, but not a lexical-flavour or numerical sequence in space. Every person, from the moment of birth, is endowed with the phenomenon of synaesthesia; however, in the process of ontogenetic development in the majority of people this synaesthesia, for physiological or psychological reasons, is diminished and not manifested in consciousness.

The inversion theory allows an explanation for the disappearance of children's synaesthesia during ontogenesis, which has a similar mechanism as children's autism. One can say that these are normal recapitulations of the pralogical consciousness of children, which are psychological embryos. In phylogenesis, human morphogenesis has outpaced psychogenesis, as proved by the findings of fossil hominins of 5-6 Mya with phenotypes close to modern man, but without signs of thought and speech at the endocranial cast. This can explain the stage-by-stage dispersion of physical and mental recapitulation in ontogenesis: anatomical signs recapitulate earlier, at the embryo stage whereas for physiological signs such as autism this occurs at the stage of infancy. Perhaps, a newborn, morphologically already a human being, from the point of view of psychological development is an embryo and if isolated from the cradle, he will be physically formed as a fully-fledged person, but mentally he will be defective; examples can be wild children. Using the idea of Jung about recapitulation of the psyche, in which in psychology is a correspondence between ontogenesis and phylogenesis, one can hypothesis that infantile synaesthesia is simply a recapitulation of earlier evolutionary stages where cross-sensory perception was the norm. Maurer & Maurer (1988) postulate that all newborns are synesthetes and an infant during postnatal development gradually shifts from cross-modal transfer founded on synesthetic confusion to cross-modal transfer founded on associations between separated sensory modalities.

Many researchers argue that the intermodality of perception and the presence in the brain of cross-modal transitions of sensory pathways is more the rule than the exception in the life of man. For example, under certain circumstances, the use of pharmaceutical drugs or hypnosis upon a person acquires synaesthesia. It resembles a situation with visualization or schizophrenia that we considered earlier.

At the initial stage of evolution, primitive man had pralogic thinking and stable cross-modal transitions of sensory pathways in the brain. In the process of psychogenesis, people learned to share modalities and independently analyse them. This process was associated with the development of thinking and developing cognitive functions influenced this process. Modern people with synaesthesia are an example of recapitulation of stable cross-modal transitions of sensory pathways that is analogous to modern carriers of pralogic thinking that recapitulated in them as atavism and which could be clinically diagnosed. With the altered state of consciousness, modern man partially lost the ability to control his perception and there is a sensory synaesthesia effect that begins to tie in with the cognitive field. These manifestations can go beyond the sensory perception and create the illusion of super-abilities, which is seen in various methods of transpersonal psychotherapy, holotropic breathing for example.

Within the framework of the endogenous theory of the origin of consciousness, when two independent brains are brought together into one organ and actively communicate and compete within it, synaesthesia was present in both hemispheres as a permanent neurophysiological reaction and ensured the formation of a complex polymodal image of the perceived object in primitive man. That enhanced survivability but also carried risk, like any game of contradictory images in the brain.

Since the human brain functions as a single organ, in the absence of coordination between the hemispheres madness arose on the cognitive level. Moreover, the synaesthesia present in different modalities and under the influence of interhemispheric misalignment, created various hallucinations that impaired reactions and orientation and ultimately reduced the survival rate of the species.

Therefore, the evolution of consciousness could only occurred by way of the rejection of synaesthesia (and according to Hegel, the negation of synaesthesia), simultaneously with the development of cognitive functions in the ancestors of man. As Homo sapiens formed, perception became more socialised and the number of synesthetic reactions decreased. The transition from one state to another occurred by the taking of a new form of what was present in a previous stage. The development of speech negated the negation of synaesthesia, and began to fulfil the role of integration between the sensory and cognitive systems of the brain. This development gradually gave way to cognitive-symbolic integration, forming metaphors and links between superficially dissimilar and unrelated things. As a consequence of this second negation, synaesthesia in modern man has been cultivated, synesthetes began to paint letters and names when reading, and music came to smell like rain or metal depending on the individual.

Metaphors

Studies of synaesthesia conducted by various researchers led to the understanding that synaesthesia gives the opportunity to associate different modalities that have similar neuropsychological mechanisms with the generation of metaphors that reveal deep connections between unrelated things. Metaphors are used for indicating bright perceptual synaesthesia as well as for specific language tropes, a description of the perception of one sensory modality in terms of another. 'Synesthetic cognition, a short-hand for the cognitive expression of synaesthesia, includes the construction and comprehension of cross-modal metaphors found in many languages' (Marks 2011, p. 58), for instance, common expressions such as 'loud colours'. The perceptual origins of synaesthesia might also serve as the origin of metaphor itself. Common language is saturated by cross-modal metaphor (Ibid.).

Conceptual metaphor expresses one thing by suggesting another as a figure of speech. It can be defined as 'the figure of speech in which a name or descriptive term is transferred to some object different from, but analogous to, that to which it is properly applicable' (The Oxford English

Dictionary, 1989). The term metaphor is ancient Greek in origin and means 'word in figurative meaning' and was included in the vocabulary of rhetoric and philosophy by the famous Athenian speaker, and teacher of eloquence, Isocrates (436-338 AD). A metaphor is a form in which the principles opposing the rules of formal, logical thinking (the law of identity, the law of prohibition of contradiction and the law of excluded middle) are embodied. A metaphor is perceived instantly, while logical operations are usually multi-stage. For instance, in order for any logical content to be perceived it often resorts to lengthy evidence. In view of the fact that the metaphor does not need any explanation or proof, one can say that it is dissociated from formal logical thinking.

Aristotle formulated the basic ideas of building a metaphor: to invent good metaphors one should notice well the similarities. 'Metaphor is the application of a word that belongs to another thing: either from genus to species, species to genus, species to species, or by analogy' (Aristotle, 2010, 1457b). He believed that the basis for the possible transfer of words from one area to another is the similarity between objects. For example, old age is as relevant to life as evening to day, so old age can be named as the evening of life. As Aristotle put it, 'It is from the metaphor that we can best get hold of something fresh' (Ibid. 1410b). However, the metaphor of Aristotle is only a substitute for words, at the level of vocabulary, and it does not affect the way of thinking about the subject.

Two thousand years later, the ideas of Aristotle were brought to perfection by British philosopher Thomas Hobbes who interpreted metaphors as a negative element unrelated to the true knowledge of the world and capable of distorting the picture of rational thinking. Hobbes (1909) supposed that taking words in their usual or most basic sense without metaphor and allegory was the exclusively applicable method for philosophical discourse and he believed that metaphors could be replaced by literal expressions without any loss of meaning. Consequently, the metaphor was understood by this scholar only as a lexical phenomenon, carrying out a number of official functions of speech. However, in 1873 Nietzsche singled out the epistemological functions of metaphor: 'truth is a mobile army of metaphors' that is the abilities to find similarities, the creative construction of a picture of the world and the ability to penetrate into the essence of complex phenomena (Nietzsche, 1964).

In modern theories, a metaphor is understood not so much as the external form of a language, but rather as the underlying structures of thought behind it. Metaphors are considered as an external manifestation of fundamental semantic processes and are derived from processes occurring in thought. For example, Richards (1936, p. 93-94) wrote that 'thought is metaphoric, and proceeds by comparison, and the metaphors of language derive there from.' When people make use of a metaphor, they 'have two thoughts of different things active together and supported by a single word, or phrase, whose meaning is a resultant of their interaction.'

Thus, the understanding of metaphor has widened from a narrowly semantic analysis of the artistic pathway to the measurement of cognitive constructs that are part of the integrative processes of subjective experience. Note in Richards' sentence: 'two thoughts of different things active together'. Working in linguistics, Richards quite unexpectedly comes to the idea that in the metaphor is interaction of two thoughts. Turning to the inversion theory of consciousness, one can say that the metaphor is formed by two different brain structures and there was a period in human anthropogenesis when all speech and thinking were woven from metaphors, generated by the competing cerebral hemispheres of primitive man.

Employing Hegel's dialectics one can say that in the evolution of the sapiens, the metaphorization of consciousness decreased (the first dialectical negation of the metaphor), due to both an increase in the dominance of one hemisphere and the development of logical functions in the ancestors of man. However, with the development of the second signal system, the development of speech, the negation of metaphor is negated and metaphor begins to fulfil the integration role between the logical and symbolic structures of consciousness, contributing to the formation of new connections in the brain. As a consequence of this second dialectical negation, metaphorical connections ensured the formation of creative functions in primitive man, which increased his survival chances.

Studies in cognitive psychology broadened the understanding of metaphor as a cognitive symbolic integration of any type, including intersensory integration. In this case, synaesthesia is a particular case of metaphor. Conversely, many authors observe the metaphor as a kind of synaesthesia. One can see the logical paradox as gracefully noted by Ramachandran & Hubbard (2001), 'The problem with the metaphor explanation is that it commits one of the classical blunders in science, trying to explain one mystery (synaesthesia) in terms of another mystery (metaphor)'. Since we know very little about the neural basis of metaphors, saying 'synaesthesia is just a metaphor' explains neither synaesthesia nor metaphor (Ibid.).

Cytowic (2002, p. 285) proposes an original approach to explain the relationship between metaphor and synaesthesia. 'Synaesthesia is not lost in the cognitive transition from infant to adult but becomes subsumed by verbal devices as cross-modal matches shift from purely sensory, to sensory-verbal, to perhaps a singularly verbal realm.' The loss of the immediacy and productivity of sensory synaesthesia during ontogenesis can be explained by symbolic manipulations originally prompted by sensory impressions. Marks (1989) concluded that perceptual experiences in meanings are multidimensional, and that verbal (semantic) knowledge taps earlier perceptual knowledge.

Despite the overflow of the ability to synaesthesia in the ability to generate metaphors in ontogenesis, metaphor's ability decreases as a person grows up. This corresponds to the studies of Gardner & Winner (1979, p. 131, 132-133), who examined the human capability to create metaphors establish that: 'The highest number of appropriate metaphors was secured from the preschool children, who even exceeded college students; moreover, these three- and four-year-olds fashioned significantly more appropriate metaphors than did children aged seven or eleven'.

Modern people who have a high level of metaphorical consciousness are examples of partial recapitulation of pralogical thinking. In this case, you can meet with both carriers of high creative abilities, who generate metaphors as from a cornucopia, and the carriers of mental pathologies. The example of young children who have a high metaphorical consciousness can serve as a confirmation of the hypothesis that infantile metaphorism is simply a repetition of the early stages of evolution, where it was the norm. The reason for a significant decrease in children's metaphorism is the same as in children's synaesthesia: these are normal recapitulations of pralogical consciousness in psychological embryos, which are children. We noted that one of the characteristic features of altered states of consciousness (ASC) is the emergence of a synesthetic experience. Similar effects occur with metaphors in ASC. Many psychological aspects of trance states such as over-associativity and short attention span, the transition from the use of verbal-logical structures to sensory (pre-verbal) structures and the presence of fragments of inner dialogue in external dialogue are ideal for generating a metaphor (Mithen 1998).

Conclusion

The three theories of human origin such as the traditional simian theory, the aquatic ape theory and the inversion theory have been examined in the book. The simian theory is based on the numerous discovered hominids fossils and it can be summarised below as follows. Around 10 Mya some species of *Dryopithecus* changed the type of feeding that resulted in a different configuration of muzzle. Their descendants developed a new mode of locomotion, morphology of the skull and post-cranial skeleton. The occurrence of *Sahelanthropus tchadensis* and its successor *Orrorin tugenensis* may represent a common ancestor of humans and apes. The first Australopithecus that evolved from mid-Miocene hominids lived in Africa and walked with shorter steps than modern Homo and their hip joints were not completely straight when walking. Also, their hands were somewhat elongated and the fingers were adapted for climbing trees.

Climate and landscape change affected the speciation of hominids and forced them to develop new ecological niches and the search for such niches inevitably led them to the ground. In particular, about 2-3 Mya the early australopithecines in Africa were replaced by gracile and robust Australopithecus and early representatives of Homo. The latest gracile *Australopithecus Garhi* found in Ethiopia dates from about 2.5 Mya. Along with the remains, primitive 'Olduvai' stone chopping tools were found. Around 2.4 Mya East African gracile Australopithecus, that most likely are the ancestors of hominids, changed their lifestyle by becoming omnivorous. As a result, the masticatory apparatus decreased in size, creating conditions for the significant enlargement of the brain that opened the way for the Hominidae family. Approximately 120 Tya in Europe and Africa, there was a new Homo that had relatively similar structure of the cerebral box as does the modern human and they differed mainly by specialized features of the facial part of the skull and the postcranial skeleton. These differences completely disappeared about 40

Tya with the appearance of modern man.

When under pressure of environmental factors in the Pliocene, African hominid ancestors moved to a mainly terrestrial life because bipedalism developed into a more optimal mode of locomotion. However, early bipedalism was not a substantial evolutionary advantage compared to other animals; it put hominid ancestors at a disadvantage in relation to predators and competitive species of monkeys and made their chances of survival very low. The supporters of the simian theory understood the weakness of their model of the bipedality, key to anthropogenesis, and sought to explain how the upright posture could arise in the face of natural selection. They proposed that culture could compensate for imperfect bipedalism.

The idea of using culture as compensation for a non-adaptive morphology was expressed in the mid-20th century by German anthropologist Rothacker, who believed that animals are creatures coherently inscribed in the environment whereas man creates for itself the environment in the form of culture. To explain where culture comes from, which is a social phenomenon not encountered elsewhere in nature, philosophical anthropology suggests using the idea of man as a special project of nature. That can be controlled by an entity such as Holy Spirit or nature that leads to the theological model of human origin. The significant difference in the evolution of hominids from other animals was their need to adapt to not only the natural surroundings but also to the constantly developing cultural environment. Consequently, anthropogenesis led to the appearance of the many behavioural and anatomical features that distinguish humans from other apes.

The inability to explain clearly the origin of bipedalism is not the only disadvantage of the simian theory. The morphological approach has revealed a discrepancy between the time of formation of the key attributes of the hominid triad and the inclusion of the labour factor in hominid evolution, which is classically taken as the cause of those attributes. There is no linkage between the beginning of use of stone tools and the development of intellectual abilities. Moreover, the nature of the use of stone tools by *Homo erectus* indicates an instinctive work rather than a rational one with an absence of progress for millions of years. Therefore, labour cannot be a causal factor of progressive evolution. Conversely, the fact that there is a well-defined advantage of speed, quality and quantity of simple labour operations from narrow specialization of labour starting as an instinctive reflex activity suggests that the first human ancestors' labour operations directed their evolution to an impasse of narrow specialization. No proponent of the labour theory has been able to explain persuasively why simian ancestors of humans would need labour as the main driver of evolution.

There are no clear solutions to simple problems such as the identity of the common ancestor of hominids and pongids from which the divergence began. There is an evolutionary time gap and no transitional links between the tailed primates and tailless apes. Researchers cannot give an evolutionary explanation for a seemingly easy matter of why apes have lost a tail, which is not a redundant organ in the tropics where there are many mosquitoes and yet monkeys have it and use it in everyday life. Also, the ancestors of chimpanzees have not been found; they seem to appear from nowhere and their ancestors cannot be identified in the fossil record. There are no clear answers to the most insistent questions of which hominid line leads to Homo sapiens or of the trigger mechanism of anthropogenesis. The simian theory of anthropogenesis came to the point where the study of particulars does not give new knowledge because there is a large gap between discoveries in palaeoanthropology and their interpretations. There is a paradox: the more researchers find fossils, the weaker their theoretical constructions become. They have too many prospective lines of human evolution that they are constantly forced to remove from the trees of anthropogenesis. However, this paleoanthropological paradox can only be resolved if one accepts that there are various evolutionary lines of human-like anthropogenesis originating from simian theory, as the found African skeletons are not human predecessors but pathologically altered skeletons of pre-sapiens who left the heavenly coast of Tethys relatively early and migrated along the river banks into equatorial Africa.

Having lost access to useful seafood, they inevitably developed a deficiency of vitamin D, which led to rickets, with symptoms such as a massive skull, thickening of the occiput, superciliary arches and curvature of extremities as observed in most findings. Deficiency of iodine caused dementia and microcephaly destroyed the endocrine system, leading to pathological changes in the bone system. Replacing the diet of marine molluscs on the river forms could lead to an abundance of various helminths and consequently to osteoporosis. Perhaps if the analysis of fossils had been done by pathologists, a better representation of the transformation of human ancestors into modern apes and into dead-end species of evolution would have resulted.

A distinct line could be built if it was a question of the development of normal anatomy, but pathological anatomy is harder to build into an evolutionary line because there are so many factors of degradation of man's distant relatives, each of them disfigured in its own way. From the coast of Tethys for millions of years came many waves of immigrants, which also degraded in turn. Africa of the anthropogenesis period is a museum of unfortunate anomalies; the remains of which researchers often declare as ancestors of man. The complexity of the proof and the weakness of criticism allow to these researchers to build any lines of anthropoid evolution and create an illusion that the mystery of the origin of man will be revealed by them in the near future.

The genesis of consciousness is the most intractable sacrament of the human mind and many modern approaches to solving the problem of the origin of human consciousness are unconvincing. Most approaches downplay the gap between the mentality of apes and human consciousness. Researchers have forgotten the rule of Conwy Lloyd Morgan (1903, p. 59): 'In no case is an animal activity to be interpreted in terms of higher psychological processes, if it can be fairly interpreted in terms of processes which stand lower in the scale of psychological evolution and development.' They widely describe animals with such concepts as thinking, intellect and cognitive behaviour. The language used in studies on abilities of animals abounds with such expressions as manipulation of a partner, group management, honest communication, deception and tolerance that researchers consider as scientific terminology. However, it does not bring them closer to an understanding of the origin of consciousness of man and usually does not increase their knowledge about animals.

In the 1920s, the English philosopher Collingwood (1946, p 248) noted 'The belief that man is the only animal that thinks at all is no doubt a superstition; but the belief that man thinks more, and more continuously and effectively, than any other animal, and is the only animal whose conduct is to any great extent determined by thought instead of by mere impulse and appetite, is probably well enough founded to justify the historian's rule of thumb'.

Modern cognitive ethology has developed these ideas and one can find examples where it is noted that a man is much inferior to jays, cedarwood, squirrels and other animals that store food in caches, judged by their ability to memorize points of terrain. A rat will beat a man in solving problems to find a way out of an intricate labyrinth and some monkeys remember faces from photos faster than people do. This is nothing surprising given that the individual mental functions in animals are formed and developed in the processes of interaction with the surrounding world; one can always find species that will significantly outperform people in a specific task. But man, using his higher mental functions, will ultimately surpass any squirrel in memorization since he will use external means of memorizing: maps, drawings or photographs or, with certain motivation, internal memory techniques based on intellectual work.

Attempting to decrease the gap between man and animal, scientists identify animals by human characteristics; thereby trying to display that there is no distinctive problem of the origin of human consciousness. Repeatedly using in their theoretical constructs such terms as possible, probably, little by little, gradually and step by step they cannot offer a real answer to the fundamental questions of how man's consciousness arose.

The second concept of human origin, which has been examined in this book, is the Aquatic Ape Theory (AAT). In contrast to the widely accepted simian theory that is based on the savannah dominant factor that made apes move to bipedalism, AAT considers the shallows of seas, rivers and lakes as the dominant crucible of human evolution. The theory presumes that an aquatic environment has acted as an important agent of selection in the evolution of humans. Consequently, many of the physical dissimilarities between human beings and apes may be explained as adaptations to wading, swimming and diving.

According to the theory, human ancestors had a semi-aquatic habitat, and only later they moved to the ground and became savanna dwellers. In the last few million years, Homo adjusted to land-dwelling life and therefore many footprints of aquatic adaptation have become moderately hidden. However, indications still exist and the AAT designates characteristics that are presented in aquatic animals and almost absent in apes and savannah mammals: enlarged pharynx, activity of sebaceous glands, paucity of apocrine glands, direction of hair follicles, a superficial fat-layer bonded to the skin, nakedness over most of the body surface, very large sebaceous glands at the hairy/naked boundaries, thermoregulatory sweating through eccrine skin glands, rich salty tears, a volitional control of breath holding and a reduced sense of smell. Supporters of the AAT believe that wading through shallow water was a major factor of the locomotion of the earliest proto-hominids that took habitats where considerable wading locomotion was favourable; the differences between the postcranial anatomy of the earliest bipedal hominids and modern humans can be explained by particular adaptations for wading or swimming. In addition, there is a connection between consumption of marine food and development of the human brain throughout all periods of human evolution.

There are many arguments both for and against this theory that are difficult to evaluate together. Many critics of the AAT note that, in comparison with the savannah theory, it does not permit the careful clarification of some inconvenient facts and it cannot give better predictions of new findings than the traditional approach. Proponents of the AAT do not develop hypothetical explanations that are tested by the traditional approaches of natural sciences. The theory has many contradictions and so does not have a wealth of testable hypotheses. The arguments of the AAT represent an unverifiable narrative of explanation of evolutionary development and demonstrate that the theory is far from the modern discussion in evolutionary biology regarding the importance of apparent adaptations.

The simian theory and the aquatic ape theory begin with eclectic descriptions of bone remains of hominids in Africa and therefore, each author builds his own evolutionary trees. De facto, many paleoanthropologists, going from ape to man, and not vice versa, created a kind of ideal, constructed primordial creature that is half-man and half-ape and called it the missing link. In appearance, it looks like the science of the past, but in essence it is a mixture of futurology and science fiction. The scientists look for characteristics of existing humans grounded upon those of the past, with the purely speculative construction that the answer to the future is in the human past.

The third explored concept, the inversion theory of human origin, employs a different methodology: to distinguish in the human body as much as possible of the anatomical and physiological exclusives that differentiate man from animals, and investigate where and what were the origins of human nature? These exclusives lead researchers to the landscape in which man was formed and the factors of evolution which worked upon the creatures from which men evolved. Past evolutionary events must be explained by causes identical to those now in operation, so the present is the key to the past. The theory takes the corporeality of modern man as the point of reference, and analyses it from an evolutionary point of view without prejudging conclusions based on external similarity.

Many human anatomical and physiological exclusives show that man's distant ancestors were marine mammals. Specifically, man is animal of marine etiology, adapted to the terrestrial way of life as opposed to aquatic apes, which are terrestrial etiology animals adapted to a semi-aquatic life. The possible landscape of anthropogenesis is the coastal border of water and land and not the tropical forest and savannah as was previously imagined.

Human progenitors could not be effective hunters or scavengers because they were in competition with species more adapted for hunting and they did not have a specific phenotype and endocrinology, which scavengers have. Therefore, the primary ecological niche of human ancestors, when they occupied its early stages, was obtaining food hiding under a hard shell. Dwelling in marine waters, they consumed mainly chiton and this niche was environmentally exclusive for a human ancestor. All animals, including human progenitors and modern men, are conservative in food preferences. Palaeolithic preferences of seafood were preserved for a very long time and scattered around the whole of Europe, from the Mediterranean to Scandinavia, are a large number of heaps of shellfish coverings, discarded by primitive people throughout the Mesolithic and Neolithic.

Based on conditions that human ancestors had to be universal creatures, able to live practically anywhere and to conduct a water-terrestrial life, the human body was formed in a relatively warm and wet atmosphere in salty water. Possibly in human evolution was a period when human development proceeded on the basis of paedomorphosis and a neotenic form of Acrodelphinidae could be an ancestor of man. Their morphology was close to their distant ancestors living at the intersection of two environments: round heads on mobile necks and limbs, as evidenced by the development of modern dolphin embryos. Neoteny in the case of archaeocetes means that their cubs were able to feed on the shore and in shallow water.

When the Tethys regression began, numerous groups of Acrodelphinidae were captured by some reservoirs without connection with the ocean due to the irregularity of the landscape. Evolution could offer to these landlocked archaeocetes only the paedomorphosis development through the neotenic form. A reduction of the water area was a positive factor for Acrodelphinidae evolution as they already had some adaptation to the shallows.

Analysis of the theories of consciousness shows that most scholars give definitions in the genetic discourse but avoid giving definitions in the essentialist discourse. Most researchers note that consciousness is an active reflection of a reality that processes external stimuli. Researchers successfully study the psyche of organisms from the simplest manifestations, for example chemotaxis, to the reflexes of higher animals. Nonetheless, they typically stop short of tackling the problem of the emergence of consciousness that in reality separates animals and man.

Some researchers find an answer in theology or esotericism that proceeds from the idea that the human consciousness is a cast of a Supreme consciousness, or His function, or His reflection. But while introducing the concept of a higher power and not explaining its genesis, they leave the logical field of academic science. Most concepts of a scientific, rather than theological, origin of consciousness believe that consciousness has an external source that it is a reflection of the surrounding world. A fundamentally different approach is possible when the cause of the origin of consciousness is communication between the different structures of one brain, so one can talk about the endogenous origin of consciousness.

The evolutionary approach considers that nature produces more and more complex forms of existence, and then it breaks down and produces something fundamentally new, and this new thing develops again into something even more complex. Thus, the brain appears: initially the nerve fibres, then the lump, next the lump is divided in two, which means the separation of functions and the possibility of competition. Nature, in general, likes to divide and collide. The partial division continues to complete separation, and at the same time each half increases and becomes more complicated autonomously. But then, in order to control the body, the interaction of the two parts of the brain begins and there is a new system of communication between the hemispheres that develops into the psychophysiological basis of thinking. Without complete separation and subsequent connection, this basis would not appear. This dialectic, the doctrine of the unity and struggle of opposites, is the essence of the inversion theory of anthropogenesis.

This division has led to the emergence of virtuality, which is a state that does not really exist but can arise under certain conditions. The laws of the existence of virtuality are not reduced to the laws of generating reality, which in itself is no longer a purely natural phenomenon. An example of virtuality is a dream. Scientists suppose with a degree of confidence that in rats, birds and other animals with a two-hemisphere brain have now been observed physiological and behavioural features similar to human dreaming. For example, Rauske, *at al.*, (2010) managed to prove that zebra finches were practising their songs whilst they slept.

The highest form of development of the divided brain is found in dolphins, in which exists the most tortuous terminal cerebrum. It is dominated by longitudinal furrows that do not extend from one hemisphere to the other so the hemispheres are not interactive to the same extent as in men. The hemispheres of the brain represent two almost autonomous brains, and this makes it possible to work alternately. A human brain is a further stage in the evolution of the brain when two independent brains are brought together into one organ and actively communicate and compete within it by using the callosal commissure. The brain had the necessary mass and morphology before the start of the formation of modern man.

This evolutionary acquisition, through which the activity of human consciousness differs from the dissociated work of the dolphin's brain, was formed as a result of the return of the Archaeocetes to the shallow sea and land. This return led to the fact that the hemispheres, practically autonomous brains, lost the ability for one to sleep at the same time as the other remained awake. As a result, they began to conflict with each other, tearing the psyche apart, which led to both the emergence of insanity and the origin of consciousness.

The original state of the psyche can be considered dual-natured; one can assume that the human consciousness is split. The unity of perception and thinking peculiar to people in their normal state is guaranteed only by the obedience of one part of the consciousness to another and by the rejection of its own initiative. At the dawn of the formation of consciousness, the normal state of hominids was not wakefulness with its wholeness of perception, but a state close to hypnotic dream. It was an internal dialogue, where a command and execution belonged to different minds living in one brain. At the stage of formation of the human brain, there was no accord, so our ancestors were schizophrenic and autistic.

Consciousness appeared in man, not because of the gradual complication of the reflex activity of ape-like ancestors but through inversion, the breakdown of animal-reflex activity through schizophrenia, the split consciousness merging due to social factors. It was necessary for people with a split mind to find agreement with other people in order to survive and not destroy each other. It can be assumed that where a society could control the behaviour of members of society, this society could survive. However, it is not necessary to represent this control as the soft control of modern society; this was a cruel and bloody control. In the primitive societies, self-healing occurred through regular, many-hour trance rituals involving dancing, cries and singing, often influenced by natural narcotic substances obtained from plants and animals.

The inversion theory supposes existing in the context of psychogenesis a recapitulation of earlier evolutionary stages of paleopsyches when there is a possibility of the appearance of rudimentary primitive consciousness, after a period of time, just as rudiments of the human body appear, for example the wisdom teeth or the accessory olfactory system. Moreover, the return of man to primitive thinking, and the slipping into primitive schizophrenia, can be much faster than the time spent in acquiring consciousness. The rejection of logical thinking for the sake of expression, the rejection of realism and unlimited expansion of self-expression are the first steps towards splitting of consciousness. Games with a changed state of consciousness are dangerous for humans, whose minds have appeared through schizophrenia.

Absolute freedom, under certain conditions, leads modern society to regress in evolutionary development by dragging it into the quagmire of primitive madness and this ultimately destroys human civilization. There were dozens of populations whose members emerged from the animal state and could have language and consciousness, but their populations have disappeared because they killed each other. They had the bonanza of evolution, they surpassed all animals and managed to survive without mind, but then they got the mind and could rule over these other animals. However, paleoanthropologists found them with skulls split by human tools. Only those whose society began to rigidly dominate and put everyone in a narrow framework survived. Only a small population managed to survive such an evolutionary bottleneck. Perhaps there were only two of them: Adam and Eve. Perhaps the passion of love become that imperative vector that allowed the split of the psyche to be overcome because man as a creature has a unique hypersexuality, characterized by the strong, all-consuming, emotional attachment to a sexual partner. When society formed, it began to dominate the psyche of individuals, cleansing it of schizophrenia.

The core idea of the inversion theory is that man achieved active reflection of external stimuli because he has an internal mirror that never, in consciousness, stops self-reflecting. Thus, the ability to reflect external incentives is an acquired characteristic of man and not the cause of the emergence of consciousness, as many philosophical schools believe. The endogenous theory of the origin of consciousness fits more organically into the scientific picture of the world and so it is possible to give an essentialist scientific definition of consciousness, defining it as the highest form of the psyche, characterized by its ability of self-reflexion. The criterion for the presence or absence of consciousness is self-awareness. The structure of self-consciousness can be characterized by how a person realizes his differences and distinguishes himself from the outside world as a subject capable of changing the surrounding reality and begins to realize his own mental processes and properties. A person who thinks he has consciousness does indeed have it, because this very thought is the criterial manifestation of consciousness. No animal, even the most intelligent, can turn the psyche upon itself. This would mean that it is capable of analysing its behaviour and its own psyche and engage in self-criticism.

The fundamental novelty of the inversion theory lies in the fact that it links morphogenesis and psychogenesis into a single entity. The rationale for morphogenesis is the foundation of the theory, but the main element is the inversion theory of psychogenesis, which states that consciousness appeared as a result of an inversion dialectical leap (discontinuity of the continuity of quantitative changes) in evolution, and not as a result of the step by step complication of monkey reflexes since the emergence of consciousness on the basis of reflexes is impossible, as proved by neuroscience during the 20th century. A possible reason for the inversion was the malfunction of the harmonious animal psyche of pre-sapiens into a schizophrenic state. The anatomical basis for this malfunction was the situation when there were two autonomous brains in one organism, which man inherited from his ancestors.

At the initial stage, an inversion was a step backwards, because pre-sapiens adaptability sharply decreased, they were on the boundary of self-destruction. However, thanks to society as the main causal factor of the occurrence of consciousness, and based on a split nature, the consciousness of Homo sapiens was formed. Numerous researchers try to build a theory of psychogenesis based on the gradual development of cognitive abilities from species to species, from ants to primates. Watching how animals get improved intellect, they transfer this approach to humans. The clarification of the problem of the beginning of consciousness should not be confined to the possibility of a direct progression from the instinctive-reflex mechanisms of the animal psyche to human consciousness and cannot be assumed to move inch by inch from the first signal system to the second signal system according to nondialectic logic.

Many philosophers see a significant drawback in the fact that the human body is not adapted to any specific landscape. However, in reality this defect is transformed into an advantage since it is not adapted well to any particular landscape, a man is paradoxically adapted to living in virtually any part of the globe. Landscape non-affiliation is in fact a huge advantage of the universal organism. All kinds of monkeys are very narrowly specialized to certain landscapes, in contrast to man. Specialization of monkeys is so narrow that gradation takes place even within a single forest. For example, the chimpanzee dwells in the second tier of the rain forest, and the gorilla lives under the crowns and they cannot switch, while people could easily do it in case of emergency. Man is not an unadapted animal; on the contrary, he is successfully embedded in the landscape of his habitat. In addition, man has a number of unique social exclusives, namely: the ability to form different types of societies, the ability to change the nature of social ties throughout life and the ability of social creativity.

The inversion theory of anthropogenesis, despite its heuristic nature, is not devoid of shortcomings. The first thing critics can point out is its eclecticism, which is a way of constructing a philosophical system by combining different positions borrowed from different philosophical and psychological systems. Every system must reckon with established facts and with true positions, no matter to what philosophical school they belong. The inversion theory of anthropogenesis tries to consider all possible directions and by way of criticism finds in them a grain of truth. Therefore, some eclecticism inherent in the theory can denote the requirement of breadth of horizon in the matter of substantiating the complex concentration of anthropogenesis and, first and foremost, psychogenesis.

Critics may also note the lack of archaeological evidence in contrast to traditional anthropology that presents thousands of archaeological finds. However, these thousands of findings, generally from barely existing or dried water basins, can confirm that there were many waves of hominoid migration from the Mediterranean area to the centre of Africa along the seacoast or large rivers such as the Nile. Many discoveries correspond more with the inversion theory of anthropogenesis than the simian approach.

Anticipating the question of why paleoanthropologists and archaeologists could not find traces of human predecessors along the banks of the Nile, it can be noted that for several millennia Homo sapiens have dug the entire Nile valley destroying the evidence of ancient cultures, so this is not a good place to make a discovery. It can be assumed that the remains of pre-sapiens that could be elements of the real human evolutionary tree are at the bottom of the Mediterranean Sea; the rise in the level of the world's oceans over the last tens of thousands of years has covered possible haunts of ancient people by a water column of over 100 meters.

Starting with the prominent works of Louis Leakey, the birthplace of man is considered to be the equatorial and subequatorial belt of the African continent as the largest number of diverse fossil hominids were found there. Moreover, in that area there is the greatest genetic diversity of man. For example Cann *et al.* (1987) note that analysis of mitochondrial DNAs, drawn from different geographic populations shows that all these stem from one woman who to has lived about 200,000 years ago, probably in Africa and apart from the African population all other the populations have multiple origins, suggesting that each area was colonised repetitively. In addition, the hypothesis of the African ancestral home of all humanity is supported by the great variety of haplogroups among the indigenous population; this result was obtained by the method of DNA chronology

The idea that the homeland of man is a region with the maximum diversity of various human species and genetic material is implicitly based on the works of the Russian botanist Vavilov, who in 1926 proposed a method for determining the ancestral home of cultural plant species according to the maximum diversity of wild-growing forms. This concept established a clear link between genetic diversity and the origin of the crop in question. However, this approach does not work with animals, because animals are mobile, unlike plants. Usually, palaeozoologists do not use it because a genomic dump does not mean the ancestral home of a particular genotype; it can be a consequence of migration.

Popper offered the theory of demarcation founded upon his perception of logical asymmetry and he said that verification is not a criterion of truth. It is logically impossible to finally verify the universal position with reference to experience. One can find in Africa 100 fossils, but one counterexample such as the finding, dated to 5.7 Mya, of footprints that represent a previously unidentified late Miocene primate that evolved human-like foot anatomy in Crete finally falsifies the corresponding universal law that that Africa gave birth to humanity. An exception, far from proving the rule, finally disproves it. If to this Crete discovery we add two finds of 2017, made in the east of the Mediterranean sea (the Misliya cave, in Israel) and in the west of the sea (Jebel Irhoud, in Morocco) when archaeologists exposed the ancient remains of *Homo sapiens* dated at around 200,000 and 300,000 years old respectively, the African concept begins to break at the seams and we should accept the Mediterranean Basin as the centre of human origin.

It is important to note that in science, in order to establish the fact of the vast mass of possible empirical observations, scientists choose only those that represent meaning within the frame of reference of a particular theory. In this context, obtaining an empirical material is an ordered sample of external reality, and scientific fact is an empirically verifiable statement, expressed in terms of a conceptual scheme and cannot exist outside the system of ideas that the author follows. If the simianists postulate that man has descended from ape-like creatures in Africa, then the collected facts will be systematized in accordance with that postulate.

The situation that has developed in the science of the origin of man (physical anthropology and palaeoanthropology) is well illustrated by the theory of Kuhn (1962) who explained the development of scientific change as the consequence of different stages of transformation paradigm. In normal science, problems are solved within the context of the leading paradigm that was simianist theory of anthropogenesis. However, in the course of time, progress in anthropology has uncovered inconsistencies, data that are not easy to interpret within the context of the current simianist paradigm. Despite the fact that these malfunctions were often resolved, they have accumulated to the point where problematic limitations of the paradigm have been exposed.

As the traditional paradigm cannot consistently account for such anomalies, physical anthropology has entered a crisis stage. In addition, one can see that this crisis cannot be resolved within the context of traditional paradigm and many significant efforts of anthropology fail. Therefore, the science undergoes a phase of scientific transformation in which the essential hypothesis of the discipline are re-examined and a new paradigm can be founded. The inversion theory of anthropogenesis will likely be a key component of the new paradigm.

Three theories of human origin were considered in this book. The inversion theory of anthropogenesis was given much attention, in comparison with the more traditional ones, due to the fact that this comparatively new theory explains many insufficiently clear problems of anthropogenesis using dialectical methodology. The scope of the book does not extend to the inversion theory in the context of the origin of language (glottogenesis). For example, the problem of how the human race found meaningful, articulate speech starting with inarticulate sounds inherited from distant ancestors. The relationship between psychogenesis and glottogenesis lies outside the scope of this book, but it presents a good opportunity for future research, not only from a narrow psycholinguistic point of view but also from a wider socio-philosophical perspective.

In addition, sociogenesis and ethnogenesis have not been analysed practically in this book but this will encourage the author to continue the study of Homo sapiens. The intention of the author is that this book will open discussion about new methodologies to explore theory of morphogenesis and theory of psychogenesis as well as providing an interesting adventure in search of human progenitors.

Acknowledgement

Numerous scholars helped to make this manuscript by sharing their knowledge in biology, philosophy, anthropology and psychology; talking scrupulously about human evolution and helping me arrange my thinking about how to get it right.

I want to thank Donald Smith, my editor, with whom I have worked on my books. His beneficial editorial experience has demonstrated invaluable in enabling a researcher to write beyond the limitations of the conventional approach. I am thankful for the substantial advice that the anonymous advisers provided me and for their aid in configuring the chapters in order for the book to take its final shape and reach completion. I would like to thank my colleagues Vadim Kalganov, Nicolas Anderson and the rest of the supportive team, who have been my day by day academic colleagues for the last years.

I am grateful to everyone at the University of Oxford that has made this book possible. I would like to make a special note to Victor Ten, who read my book with admirable care and attention, making some especially valuable suggestions that I have incorporated. Without his extraordinary philosophical ideas, this book could not be born.

I would especially like to show appreciation to my old friend John Peterson who helped me in the collection of information for my research. I am also deeply grateful to the people who took many hours to talk with me about palaeoanthropology, archaeology and philosophy; their ideas contributed to the creation of my vision of anthropogenesis: Denis

Kondratiev and Mrs Demidova.

I would like to express my special appreciation to Philip Cox for helping me, his attention to nuances were astonishing, he read every draft of this book and made every chapter better. Thank you to everyone around Russia and the United Kingdom who have invited me to speak and contribute articles over the past few years; you know who you are.

Last, but undoubtedly not least, I want to thank my family for their love and support. Most of all, unquestionably, I must thank my son Mikhail who could not play tennis with me as often as he wanted due to my writing of the book and my wife, Maria, who surrounded me with love and assistance. I deeply appreciate her patience as she endured my writing.

Mikhail Glazunov

Bedfordshire, England

June 2019

Bibliography

Agust, J., & Antón, M. (2002). *Mammoths, Sabertooths, and Hominids: 65 Million Years of Mammalian Evolution in Europe.* New York: Columbia University Press.

Aiello, L. C. (2010). Five years of Homo floresiensis. *American Journal of Physical Anthropology, 142*(2), 167–179.

Aristotle. (2010). *Rhetoric.* New York: Cosimoclassics.

Armstrong, J., & Malacinski, G. (1989). *Developmental Biology of the Axolotl.* Oxford University Press. Retrieved from The Encyclopedia of Life (EOL): http://eol.org/pages/1019571/details

Asfaw, B. W. (1999). Australopithecus garhi: a new species of early hominid from Ethiopia. *Science, 284*, 629-635.

Bekhtereva, N. P. (1978). *The Neurophysiological Aspects of Human Mental Activity.* Oxford University Press .

Berger Lee, Darryl John, Ruiter J, Churchil Steven. (2015, September). Homo naledi, a new species of the genus Homo from the Dinaledi Chamber, South Africa. *eLife.*

Berger, L. R., de Ruiter, D. J., Churchill, S. E., Schmid, P., Carlson, K. J., Dirks, P. H., & Kibii, J. M. (2010). Australopithecus sediba: a new species of Homo-like australopith from South Africa. *Science, 328*, 195–204.

Berne, E. (1964). *Games People Play – The Basic Hand Book of Transactional Analysis.* New York: Ballantine Books.

Bleuler, E. P. (1950). Dementia Praecox of the Group of Schizophrenias . New York: International Universities Press.

Blizniak, D. (2012). *The Aquatic Ape Theory.* Retrieved May 1, 2017, from http://channelingerik.com: http://channelingerik.com/the-aquatic-ape-theory/

Bolk, L. (1926). *Das Problem der Menschwerdung.* Jena.

Brain, C. (2013). *Raymond Dart and our African origins. A Century of Nature.* University of Chicago Press.

Bricoleur. (2015). Trance States and Metaphor Generation. *Journal of Visionary Art, Sacred art, Traditionalism and Esoteric*

Studies. Retrieved December 19, 2017, from http://sophiaimaginalis.com/articles/trance-states-and-metaphor-generation/

Broom, R. (1950). *Finding the missing link* . Watts.

Brunet, M., Guy, D., P., H.T., M., & Likius A., A. D. (2002, Aug 15). A new hominid from the Upper Miocene of Chad, Central Africa. *Nature, 418(6899)*, 145-51.

Bunn, H. (2007). *Evolution of the Human Diet.* New York: Oxford University Press.

Caldwell, D., & Caldwell, M. (1972). *The World of the Bottlenosed Dolphin.* Philadelphia: Lippincott Co.

Cann, R., Stoneking, M., & Wilson, A. (1987). Mitochondrial DNA and human evolution. *Nature*(325), 31-36.

Carroll, R. (1988). *Vertebrate Paleontology and Evolution.* New York: W. H. Freeman and Company.

Cartmill, M., Smith, F. H., & Brown, K. B. (2009). *The Human Lineage (Foundation of Human Biology)* . Wiley-Blackwell.

Chertok, L. (1982). *Unknown in the human psyche (Nepoznannoye v psikhike cheloveka)*. Moscow: Progress.

Cohen, C., & Hublin, J. J. (2017). *Boucher de Perthes - Les origines romantiques de la préhistoire.* BELIN.

Collingwood, R. G. (1946). *The Idea of History.* Oxford: Oxford University Press.

Crow, T. J. (1997, Dec 19). Is schizophrenia the price that Homo sapiens pays for language? *Schizophr Res., 28(2-3)*, 127-41.

Cunnane, S. C. (2005). *Survival of the Fattest. The Key to Human Brain Evolution.* World Scientific Publishing Co Pte Ltd.

Cytowic, R. E. (1993). *The Man Who Tasted Shapes.* New York: Tarcher / Putnam.

Cytowic, R. E. (2002). *Synesthesia: A Union of the Senses - Second Edition.* A Bradford Book.

Daisuke, K., Koho, R. T., & Kaifu, Y. (2013). Brain size of Homo floresiensis and its evolutionary implications.

Dalrymple, G. B., & Lanphere, M. (1970). *Potassium-argon dating. Principles, Techniques and Applications to Geochronology.* San Francisco: W. H. Freeman and Co.

Dalton, R. (2010). Africa's next top hominid Ancient human relative could walk upright. *Nature.* Retrieved May 1, 2017, from

http://www.nature.com/news/2010/100621/full/news.2010.3
05.html

Danilova, E. (1965). *Evolution of the hand in connection with the issues of anthropogenesis (Evolyutsiya ruki v svyazi s voprosami antropogeneza)*. Kiev: Naukova Dumka.

Dart, R. A. (1959). *Adventures with the Missing Link.* New York: Harper & Brothers .

Dart, R., & Craig, D. (1959). *Adventures with the Missing Link.* New York: Harper & Brothers.

Dean, F. (2000). *Primate diversity.* New York: W. W. Norton & Company.

Descartes, R. (1911). Meditations on First Philosophy (1641). In *The Philosophical Works of Descartes.* Cambridge: Cambridge University Press. Retrieved from http://selfpace.uconn.edu/class/percep/DescartesMeditations.pdf

Dollo, L. (1893). *Les Lois De L'évolution.* New York: Arno.

Dowson, T. A. (1994). Reading art, writing history: rock art and social change in southern Africa. *World Archaeology, 25*(3), 332-345.

Dronfield, J. (1993). Ways of seeing, ways of telling: Irish passage-tomb art, style and the universality of vision. In L. M. P., *Rock Art Studies: the Post-Stylistic Era* (pp. 179–93). Oxford: Oxbow.

Dunbar, R. I. (1988). Theory of mind and the evolution of language. In M. S.-K. J. Hurford, *Approaches to the Evolution of Language* (pp. 92–110). Cambridge: Cambridge University Press.

Dunbar, R. I. (2004). *The Human Story*. London: Faber and Faber.

Durkheim, É. (1922). *Éducation et sociologie.* Paris: Félix Alcan.

Ellis, V. (1987). Swimming monkeys and apes - know their biology. *Proceedings Western Regional Meeting.* Fresno CA: AAZPA.

Ellis, V. D. (1986). Proboscis monkey and aquatic ape. *Sarawak Mus J., 36*, 251-262.

Engels, F. (1934). *The Part played by Labour in the Transition from Ape to Man*. Moscow: Progress Publishers.

Engels, F. (1954). *The Dialectics of Nature*. Moscow: Progress.

Falk, D. (2011). *The Fossil Chronicles: How Two Controversial Discoveries Changed Our View of Human Evolution*. University of California Press.

Fischman, J. (2005). Family ties: Dmanisi find. *National Geographic*, 17-27.

Foley, R. (1987). *Another Unique Species: Patterns in Human Evolutionary Ecology*. London: Longman.

Fossilworks. (2017). *Champsodelphis Gervais 1848 (whale)*. (P. Database, Producer) Retrieved November 10, 2017, from Fossilworks and the Paleobiology Database: http://fossilworks.org/bridge.pl?a=taxonInfo&taxon_no=366 53

Freud, S. (1961). *Beyond the Pleasure Principle*. New York – London: W. W. Norton & Company, Inc.

Frith, C., & Johnstone, E. C. (2003). *Schizophrenia: A Very Short Introduction*. OUP Oxford.

Gardner, H., & Winner, E. (1979). The Development of Metaphoric Competence: Implications for Humanistic Disciplines. In S. Sacks, *On Metaphor*. University of Chicago Press.

Garstang, W. (1922). The Theory of Recapitulation: A Critical Restatement of the Biogenetic Law. *Journal of the Linnean Society (Zoology), 35*, 81-101.

Garwin, L., & Lincoln, T. (2003). *A Century of Nature: Twenty-One Discoveries that Changed Science and the World*. University of Chicago Press.

Gee, H. (2000). *Search of Deep Time: Beyond the Fossil Record to a New History of Life*. Cornell University Press.

Gehlen, A. (1988). *Man: His Nature and Place in the World*. Columbia University Press.

Gerard D.Gierliński, G. N. (2017, October). Possible hominin footprints from the late Miocene (c. 5.7 Ma) of Crete? *Proceedings of the Geologists Association, 128*(5-6), 697-710. Retrieved from https://www.researchgate.net/publication/319411867_Possib

le_hominin_footprints_from_the_late_Miocene_c_57_Ma_o
f_Crete

Gibbons, A. (2007). *The First Human: The Race to Discover Our Earliest Ancestors.* Anchor Books.

Gingerich, P. D., & Russell, D. E. (1981). Pakicetus inachus, A New Archaeocete (Mammalia, Cetacea) from the Early-Middle Eocene Kuldana Formation of Kohat (Pakistan). *Contributions from the Museum of Paleontology, The University of Michigan, 25*, 235-246.

Gingerich, P. D., Wells, N. A., Russell, D. E., & Shah, S. M. (1983). Origin of Whales in Epicontinental Remnant Seas: New Evidence from the Early Eocene of Pakistan. *Science, 220(4595)*, 403–406.

Golovanova, L. V. (2010). Significance of Ecological Factors in the Middle to Upper Paleolithic Transition. *Current Anthropology*(51), pp. 655-691.

Gould, S. J. (1985). *Ontogeny and Phylogeny.* Belknap Press of Harvard University Press.

Gregory, W. K. (1920). On the structure and relation of Notharctus, an American Eocene primate. *Mem. Am. Mus. Nat. Hist, 351*, 243.

Haeckel, E. H. (2010). *The Evolution of Man.* Nabu Press.

Haile-Selassie Y, S. B. (2010). New hominid fossils from Woranso-Mille (Central Afar, Ethiopia) and taxonomy of early Australopithecus. *Am J Phys Anthropol., 141(3)*, 406-17.

Hardy, A. (1960). Was man more aquatic in the past?. *New Scientist, 7*, 642–645.

Hardy, A. C. (1977). Was there a Homo aquaticus? *Zenith, 15 (1)*, 4–6.

Hershkovitz I., W. G. (2018). The earliest modern humans outside Africa. *Science, 359*(6374), 456-459.

Hilgard, E. (1991). A neodissociation interpretation of hypnosis. In J. R. Ed. by S.J. Lynn, *Theories of hypnosis: Current models and perspectives* (pp. 83—101). New York: Guilford Press.

Hirsch, S. R., & Weinberger, D. R. (2003). *Schizophrenia.* Wiley-Blackwell.

Hobbes, T. (1909). *Leviathan (reprinted from the edition of 1651).* Oxford : Clarendon Press .

Hothersall, D. (2003). *History of Psychology, 4th Edition.* McGraw-Hill Education.

Hublin, J.-J., & Ben-Ncer, A. B. (2017). New fossils from Jebel Irhoud, Morocco and the pan-African origin of Homo sapiens. *Nature*(546), 289–292. Retrieved from https://www.nature.com/news/oldest-homo-sapiens-fossil-claim-rewrites-our-species-history-1.22114

Huizinga, J. (1949). *Homo Ludens a Study of the Play-Element in Culture.* London, Boston and Henley: Routledge & Kegan Paul.

Hunt, K. D. (1996). The postural feeding hypothesis: an ecological model for the origin of bipedalism. *S Afr J Sci , 9,* 77–90.

Huxley, T. H., & Youmans, W. J. (1869). *The Elements of Physiology and Hygiene: A Text-book for Educational Institutions.* New York: D. Appleton.

Isaac, B. (1989). *The Archaeology of Human Origins: Papers by Glynn Isaac.* Cambridge: Cambridge University Press.

Jacobi, R. H. (2008). The 'Red Lady' ages gracefully: New Ultrafiltration AMS determinations from Paviland. *Journal of Human Evolution*(55), 898-907.

Johanson, D. C., & Wong, K. (2009). *Lucy's Legacy: The Quest for Human Origins.* Harmony Books.

Johanson, D., & Edgar, B. (1996). *From Lucy to Language.* New York: Simon and Schuster.

Johanson, D., & Shreeve, J. (1989). *Lucy's Child: The Discovery of a Human Ancestor.* London: Viking.

Jung, C. (1970). Archaic man . In *The Collected Works of C. G. Jung* (Vol. 10). Princeton: Princeton University Press.

Kate, W. (2015). Mysterious New Human Species Emerges from Heap of Fossils. *Scientific American.* Retrieved May 1, 2017, from https://www.scientificamerican.com/article/mysterious-new-human-species-emerges-from-heap-of-fossils/

Keith, A. (1948). *A New Theory of Human Evolution.* London: Watts & Co.

Kelley, J. (2002). The hominoid radiation in Asia. In W. (. Hartwig, *The primate fossil record* (pp. 369–384). Cambridge University Press.

Khrisanfova, E., & Perevozchikov, I. (2002). *Anthropology*. Higher School, Moscow State University.

Kraepelin, E. (1919). *Dementia Praecox and Paraphrenia*. Edinburgh: ES Livingstone.

Kretschmer, E. (1922). *Medizinische Psychologie*. Leipzig: Thieme.

Kubo, D., Kono, R., & Kaifu, Y. (2013). Brain size of Homo floresiensis and its evolutionary implications. *Proceedings: Biological Sciences, 280(1760)*, 1-8. Retrieved from http://www.jstor.org/stable/23478589

Kuhn, T. (1962). *The Structure of Scientific Revolutions* . Chicago: The University of Chicago Press.

Kuhn, T. (n.d.). *The Structure of Scientific Revolutions* . Chicago: The University of Chicago Press.

Kuliukas, A. (2011 a). A Wading Component in the Origin of Hominin Bipedalism. In M. V. Vaneechoutte, *Was Man More Aquatic In The Past?* (pp. 36-66). Basel: Bentham.

Kuliukas, A. (2011). Langdon's Critique of the Aquatic Ape Hypothesis: It's Final Refutation,. In K. A. Verhaegen M., *Was Man More Aquatic In The Past? Fifty Years After Alister Hardy: Waterside Hypothesis Of* (pp. 213-225). Basel: Bentham.

Kuliukas, A. (2014). Removing the "hermetic seal" from the Aquatic Ape Hypothesis: Waterside. *Advances in Anthropology, 4*, 64-167.

Kuliukas, A. (2016). *A wading component in the origin of hominid bipedality? Ph.D. Thesis.* University of Western Australia.

Kuzmin, A. (1976). *Structure and formation of the skull of cetaceans in ontogenesis (Stroyeniye i formirovaniye kostnogo cherepa kitoobraznykh v ontogeneze. Avtoreferat dissertatsii).* Vladivostok, Russia: Author's abstract of the dissertation.

Lamarck, J. B. (1809). *Philosophie Zoologique.* Editions Flammarion.

Laming, A. (1959). *Lascaux: Paintings and Engravings.* Harmondsworth: Penguin Books.

Langdon, J. H. (1997). Umbrella hypotheses and parsimony in human evolution: a critique of the Aquatic Ape Hypothesis. *Journal of Human Evolution, 33*, 479-494.

Leakey Meave G., F. S. (2012, August). New fossils from Koobi Fora in northern Kenya confirm taxonomic diversity in early Homo. *Nature*(488), 201–204.

Leakey, L., Tobias, P. V., & Napier, J. R. (1964). A New Species of the Genus Homo from Olduvai Gorge. *Nature, 202*(4927), 7-9.

Leakey, M. D. (1984). *Disclosing the Past.* Doubleday.

Leakey, M. G. (2001). New hominin genus from eastern Africa shows diverse middle Pliocene lineages. *Nature*(410), 433 - 440 .

Leakey, M. G., Feibel, C. S., & McDougall, I. &. (1995). New four-million-year-old hominid species from Kanapoi and Allia Bay, Kenya. *Nature , 376*, 565–571.

Leakey, R. E. (1973). Evidence for an advanced Plio-Pleistocene hominid from East Rudolf, Kenya. *Nature, 242*, 447-450.

Leakey, R., & Walker, A. (1985). Homo Erectus Unearthed: A Fossil Skelton 1,600,000 Years Old. *National Geographic, 168*, 624-629.

Leontyev, A. N. (1977). Activity and Consciousness. In *Philosophy in the USSR, Problems of Dialectical Materialism* (pp. 180-202). Moscow. Retrieved May 1, 2017, from https://www.marxists.org/archive/leontev/works/1978/ch4.htm

Leontyev, A. N. (1978). *Activity and Consciousness.* Marxists Internet Archive.

Leontyev, A. N. (1981). *Problems of the Development of the Mind.* Moscow: Progress.

Leroi-Gourhan, A. (1982). *The Dawn of European Art: An Introduction to Palaeolithic Cave Painting .* Cambridge University Press .

Lévy-Brɹhl. (1985). *How Natives Think (1910 in French).* (L. A. Clare, Trans.) Princeton: Princeton University Press.

Lévy-Brɹhl, L. (1923). *Primitive Mentality.* London: Allen & Unwin.

Lewis-Williams, J. D. (2002). A cosmos in stone: interpreting religion and society through rock art. CA: Walnut Creek.

Lewis-Williams, J. D., & Dowson, T. A. (1988, April). The Signs of All Times: Entoptic Phenomena in Upper Paleolithic Art. *Current Anthropology, 29*(2), 224.

Lieberman, D. E. (2001). Another face in our family tree. *NATURE, 410*, 419-420.

Lindblad, J. (1987). *Människan : du, jag - och den ursprungliga.* Stockholm: Bonnier.

Locke, J. (1979). *An Essay Concerning Human Understanding.* Oxford University Press.

Lordkipanidze, D. T., Jashashvili, A., Vekua, M. S., Ponce de León, C. P., Zollikofer, G. P., Rightmire, H., . . . A. Mouskhelishvili, M. (2007). Postcranial evidence from early Homo from Dmanisi, Georgia. *Nature, 449*, 305–310.

Lovejoy, C. O. (1981). The origin of man. *Science, 211*, 344–350.

Lovejoy, C. O. (1988). Evolution of Human walking. *Scientific American, 259 (5)*, 82–89.

Lovejoy, C. O. (2009, October 02). Careful Climbing in the Miocene: The Forelimbs of Ardipithecus ramidus and Humans Are Primitive. *Science, 326*(5949), 70-70e8.

Luria, A. (1968). *The Mind of a Mnemonist.* New York: Basic Books.

Marks, L. (2011). Synesthesia, Then and Now. *Intellectica, 55*, 47-80.

Marks, L. E. (1989). On cross-modal similarity: The perceptual structure of pitch, loudness, and brightness. *Journal of Experimental Psychology: Human Perception and Performance, 15*, 586-602.

Marx, K. (1845 (1969),). Theses On Feuerbach. In E. F, *First Published: As an appendix to Ludwig Feuerbach and the End of Classical German Philosophy in 1888* (pp. 13-15). Moscow: Progress . Retrieved May 1, 2017, from https://www.marxists.org/archive/marx/works/1845/theses/

Maurer, D., & Maurer, C. (1988). *The World of the Newborn.* New York: Basic Books.

McCown, T. D., & Keith, A. (1939). *The stone age of Mount Carmel: The fossil human remains from the Levalloiso-Mousterian.* Oxford: Clarendon Press.

Mchedlidze, G. (1976). *The main features of the paleobiological history of cetaceans (Osnovnyye cherty paleobiologicheskoy istorii kitoobraznykh).* Tbilisi.

Meave G. Leakey, Fred Spoor, Frank H. Brown, Patrick N. Gathogo, Christopher Kiarie, Louise N. Leakey & Ian McDougall. (2001). New hominin genus from eastern Africa shows diverse middle Pliocene lineages. *Nature*(410), 433–440.

Mithen, S. (1998). *The Prehistory Of The Mind: A Search for the Origins of Art, Religion and Science.* W&N.

Mithen, S. J. (2006). The evolution of social information transmission in Homo. In J. S. Wells, *Social Information Transmission and Human Biology* (pp. 151-170). London: CRC Press.

Mongait, A. (1973). *Archeology of Western Europe. The Stone Age.(Arkheologiya Zapadnoy Yevropy. Kamennyy vek)* (Vol. 1). Moscow: Nauka.

Montagna, W. (1985). The evolution of human skin. *J. Hum. Evol., 14*, 3-22.

Montagu, A. (1955). Time, Morphology, and Neoteny in the Evolution of Man. *American Anthropologist, 57*, 13-17.

Moore, J. (2003). *Aquatic Ape Theory: Sink or Swim?* (J. Moore, Producer) Retrieved May 1, 2017, from Aquatic Ape Theory: Sink or Swim?: http://www.aquaticape.org

Morgan, C. L. (1903). *An introduction to comparative psychology, 2nd edition.* London: W. Scott.

Morgan, E. (1972). *The Descent of Woman.* Souvenir Press.

Morgan, E. (1997). *The Aquatic Ape Hypothesis.* London: Souvenir Press.

Morris, D. (1967). *The Naked Ape.* London: Jonathan Cape.

Murray, J., Peter, N. H., Seoighe, C., McCormack, G. P., Williams, D. M., & Harper, D. A. (2015). The Contribution of William

King to the Early Development of Palaeoanthropology. *Irish Journal of Earth Sciences, 33*, 1-16.

Nadia, D. (2015, September 15). Mystery Lingers Over Ritual Behavior of New Human Ancestor. *National Geographic.*

Nasrallah HA, W. D. (1986). *Handbook of Schizophrenia.* Amsterdam: Amsterdam, Elsevier Science Publishers.

Neumann, E. (2015). *The Great Mother: An Analysis of the Archetype.* Princeton University Press.

Nietzsche, F. (1964). On Truth and Lie in the Extra-Moral Sense. In *The Complete Works of Friedrich Nietzsche* (M. Miigge, Trans.). New York: Russell and Russell.

Orne, M. T. (1979). On the simulating subject as a quasi-control group in hypnosis research: What, why, and how. In E. F. (Eds.), *Hypnosis: Developments in research and new perspectives* (pp. 519 –565). New York: Aldine.

Päabo, S. (2014). *Neanderthal Man: In Search of Lost Genomes.* Basic Civitas Books.

Palmer BA, P. V. (2005). The lifetime risk of suicide in schizophrenia: a reexamination. *Arch Gen Psychiatry, 62(3)*, 247-53.

Pavlov, I. P. (1955). *Selected works.* Moscow: Foreign Languages Publishing House.

Pond, C. M. (1987). Fat and Figures. *New Scientist, 4*, 62-66.

Porshnev, B. (1974). *On the beginning of human history (the problem of paleopsychology) (O nachale chelovecheskoy istorii (problemy paleopsikhologii)).* Moscow: Mysl.

Priest, S. (1991). *Theories of the Mind. A compelling investigation into the ideas of leading philosophers on the nature of the mind and its relation to the body.* The Penguin Books.

Pull, C. B. (2002). Diagnosis of Schizophrenia: A review. In S. N. Maj M, *Schizophrenia. 2nd ed.* (pp. 1-37). Chichester, UK: John Wiley & Sons.

Ramachandran, V., & Hubbard, E. (2001). Synaesthesia—A Window Into Perception, Thought and Language. *Journal of Consciousness Studies, 8*, 3–34.

Rauske P., C. Z. (2010). Neuronal Stability and Drift across Periods of Sleep: Premotor Activity Patterns in a Vocal Control Nucleus of Adult Zebra Finches. *J Neurosci, 30(7)*, 2783–2794.

Reed, K. (1997). Early hominid evolution and ecological change through the African Plio-Pleistocene. *J Hum Evol., 32*, 289–322.

Reynolds, V. (1991). Cold and watery? Hot and dusty? Our ancestral environment and our ancestors themselves: An overview. In W. J. Roede M, *The aquatic ape: Fact or fiction?* (pp. 331-41). London: Souvenir Press.

Reynolds, V. (1991). Cold and Watery? Hot and Dusty? Our Ancestral Environment and Our Ancestors Themselves: an Overview. In M. Roede, J. Wind, J. Patrick, & V. e. Reynolds, *Aquatic Ape: Fact of Fiction Proceedings from the Valkenburg Conference* (p. 340). London: Souvenir Press.

Richards, I. A. (1936). *The Philosophy of Rhetoric.* Oxford: Oxford University Press.

Richards, R. J. (2008). *The Tragic Sense of Life: Ernst Haeckel and the Struggle over Evolutionary Thought.* University of Chicago Press.

Rightmire, G. P. (1993). *The Evolution of Homo Erectus: Comparative Anatomical Studies of an Extinct Human Species.* Cambridge University Press.

Robinson, J. T. (1972). *Early Hominid Posture and Locomotion.* Chicago: University of Chicago Press.

Rodman, P., & McHenry, H. (1980). Bioenergetics and the origin of hominid bipedalism. *Am J Phys Anthropol , 52*, 103–106.

Roggenbuck, M., & Ida Bærholm Schnell, N. B. (2014). The microbiome of New World vultures. *Nature Communications, 5 (5498).* Retrieved May 1, 2017, from https://www.nature.com/articles/ncomms6498

Rose, M. D. (1991). The process of bipedalization in hominids. In S. B. Coppens Y, *Origine(s) de la bipe'die chez les hominide's.* Paris: CNRS.

Sawyer, G., Deak, V., Sarmiento, E., & Milner, R. (2007). *The Last Human: A Guide to Twenty-Two Species of Extinct Humans.* New Haven: Yale University Press.

Scheler, M. (1928). *Die Stellung des Menschen im Kosmos.* Retrieved March 2017, from https://www.scribd.com/doc/97357646/Max-Scheler-1928-Die-Stellung-Des-Menschen-Im-Kosmos

Schneider, K. (1959). *Clinical Psychopathology.* (M. Hamilton, Trans.) New York: Grune & Stratton, Inc.

Schultz, A. H. (1960). *Age changes in primates and their modification in man.* Oxford: Pergamon Press.

Sechenov, I. (1866). *Reflexes of the brain (Refleksy golovnogo mozga).* St. Petersburg: AST Publishing House, (2014) Electronic version of the book, https://www.litres.ru/ivan-sechenov/refleksy-golovnogo-mozga-2/.

Senut, B., Pickford, M., & Dominique Gommery, P. M. (2001). First hominid from the Miocene (Lukeino Formation, Kenya). *Earth and Planetary Sciences, 332,* 137–144.

Shevchenko, Y. (1971). *Evolution of the cerebral cortex of primates and humans (Evolyutsiya kory mozga primatov i cheloveka)*. Moscow: Moscow state university.

Shipman, P. (2001). *The man who found the missing link: Eugène Dubois and his lifelong quest to prove Darwin right*. Simon & Schuster.

Simpson AI, M. B.-W. (2004). Homicide and mental illness in New Zealand, 1970-2000. *Simpson AI(1), McKenna B, Moskowitz A, Skipworth J, Barry-Walsh J., 185*, 394-8.

Simpson, G. G. (1951). *Horses: The Story of the Horse Family in the Modern World and Through Sixty Million Years of History*. Oxford University Press.

Sperry, R. W. (1961). Cerebral Organization and Behavior: The split brain behaves in many respects like two separate brains, providing new research possibilities. *Science, 133 (3466)*, 1749–1757.

Spirkin, A. (1983). *Dialectical Materialism*. Progress Publishers.

Spitsyn, V. (1989). Modern ideas about the evolution of the order of primates in the light of the data of molecular biology

(Sovremennyye predstavleniya ob evolyutsii otryada primatov v svete dannykh molekulyarnoy biologii). In *Biological Evolution and Man (Biologicheskaya evolyutsiya i chelovek)*. Moscow.

Stedman, H., & Kozyak B, W. N.-P. (2004, Mar 25). Myosin gene mutation correlates with anatomical changes in the human lineage. *Nature, 428(6981)*, 415-8.

Suryani, L., & Jensen, G. (1993). *Trance and Possession in Bali: Window on Western Multiple Personality, Possession Disorder and Suicide*. Kuala Lumpur, New York: Oxford University Press.

Suwa, G., Asfaw, B., Kono, R. T., Kubo, D., Lovejoy, C. O., & White, T. D. (2009). The Ardipithecus ramidus Skull and Its Implications for Hominid Origins. *Science, 326*(5949), 68-68.

Suwa, G., Kono, R. T., Simpson, S. W., Asfaw, B., Lovejoy, C. O., White, T. D., & al., e. (2009). Paleobiological implications of the Ardipithecus ramidus dentition. *Science*(326), 94-9.

Swisher, C. C., Curtis, G. H., & Lewin, R. (2000). *Java Man: How Two Geologists Changed Our Understanding of Human Evolution*. Chicago: University of Chicago Press.

Ten, V. (2005). *'... From the sea foam'. Inversion theory of anthropogenesis ("...Iz peny morskoy". Inversionnaya teoriya antropogeneza)*. Kaluga, Russia.

Ten, V. (2013). *Farewell, monkey The truth about the origin of man (Proshchay, obez'yana. Pravda o vozniknovenii cheloveka)*. St Petersburg: Peterburgskiy modnyy bazar.

Ten, Victor. (2011). *Archeology of man. The origin of the body, mind, language (Arkheologiya cheloveka. Proiskhozhdeniye tela, razuma, yazyka)*. Russia: Kiryanov Publishing House.

The Oxford English Dictionary. (1989). *Metaphor* (Vol. 9). Clarendon.

Tim D. White, B. A.-S. (2009, October). Ardipithecus ramidus and the Paleobiology of Early Hominids. *Science, 326*(5949), 64-86.

Ukhtomsky, A. (2002). *The Dominant*. St-Petersburg: Classics of Psychology, Piter Series, In Russian.

Van Couvering, J. A. (2000). The Pliocene. In I. T. Eric Delson, *Encyclopedia of Human Evolution and Prehistory: Second Edition* (pp. 574–576). New York: Routledge.

Vasilevskaya, G. (1984). Thoracic limb of modern dolphins in evolutionary illumination (Grudnaya konechnost' sovremennykh del'finov v evolyutsionnom osveshchenii). In *Macroevolution (Makroevolyutsiya)*. Moscow.

Verhaegen, M. (1993). Aquatic versus savanna: comparative and paleo-environmental evidence. *Nutrition and Health, 9*, 165-191.

Verhaegen, M., & Stephen Munro, M. V.-O. (2007). The original econiche of the genus Homo. Open plain or waterside? In S. I. (Editor), *Ecology Research Progress* (pp. 155-186). Nova Science Publishers, Inc.

Verhaegen, M., Puech, F., & & Munro, S. (2002). Aquarboreal ancestors? *Trends in Ecology & Evolution, 17*, 212-217.

Vygotsky, L. S. (1978). *Mind in Society Development of Higher Psychological Processes*. Cambridge, Massachusetts: Harvard University Press New edition.

Ward, C., Leakey, M., & Walker, A. (1999, May 11). The new hominid species Australopithecus anamensis. *Evolutionary Anthropology: Issues, News, and Reviews, 7*(6), 197–205.

Washburn S. L. (1960). Tools and human evolution. *Scientific American, 203 (3)*, 63–75.

Washburn, S. L. (1967). Behavior and the origin of man. *Proceedings of the Royal Anthropological Institute of. Great Britain and Ireland, 3*, 21-27.

Westenhöfer, M. (1942). *Der Eigenweg des Menschen.* Berlin: Mannstaedt & Co.

Wheeler, P. (1992). The thermoregulatory advantages of larger body size forhominids foraging in savannah environments. *J Hum Evol* (23), 351–362.

Wheller, P. (1984). The Evolution of Bipedality and Loss of Functional Body Hair in Hominoids. *Journal of Human Evolution, 13*, 91-98.

William A. Haviland, D. W. (2010). *Evolution and Prehistory: The Human Challenge.* Wadsworth Publishing.

Wolpoff, M. H. (1999). *Paleoanthropology.* Boston Mass: McGraw-Hill.

Wolpoff, M. H., Hawks, J., Senut, B., & Pickford, M. a. (2006). An ape or the ape: Is the Toumaï Cranium TM 266 a Hominid? *PaleoAnthropology*, 36–50.

Wood, J. F. (1929). *Man's Place Among the Mammals.* London: Edward Arnold.

Wortman, J. (1903). Studies of Eocene Mammalia in the Marsh collection, Peabody Museum. Part II, primates. *Am. J. Sci., 15*, 163–176, 399–414, 419–436.

Zaichenko, A., Anisimova, E., & Nikolenko, V. (2002). Philo-ontogenetic typology of the human brain's skull (Filo-ontogeneticheskaya tipologiya mozgovogo cherepa cheloveka). *IV International Congress on integrative anthropology* (pp. 140-141). St. Petersburg: Publishing house SPBGMU (Russian).

Zaichenko, A., Anisimova, E., & Speransky, V. (2002). Gracilization: the main trend of transformation of the structure of the bones of the cerebral cranium in anthropogenesis (Gratsilizatsiya: osnovnaya tendentsiya preobrazovaniy struktury kostey mozgovogo cherepa v antropogeneze). *Congress on integrative anthropology* (pp. 139-140). St. Petersburg: Publishing house SPBGMU (Russian).

Zalmout, I. S. (2010). New Oligocene primate from Saudi Arabia and the divergence of apes and Old World monkeys. *Nature, 466*, 360–364.

Zilberman U, S. P. (2004, Jun). Evidence of amelogenesis imperfecta in an early African Homo erectus. *J Hum Evol., 46(6)*, 647-53.

N

necrophagia 161

Negroid people 166

neoteny 152

Ngeneo Bernard 38

Nietzsche 247

O

Old World monkey 130

Olduvai Gorge 34, 35

Omomyidae 130

Oreopithecus 131

Orrorin 43, 47, 48, 55, 96, 97

Ouranopithecus 132, 142

P

paedomorphosis 13, 120, 149, 150, 151, 153, 155, 157, 158, 159, 258

paedophilia 157, 158

Pakicetus 137

Pakistan 137

Paleoanthropus 212, 213

parabiosis 181

Paranthropus 31, 32, 47, 76

Paranthropus robustus 31, 32

Pavlov Ivan 181, 185

Pierolapithecus catalaunicus 142

Pithecanthropus 27

Plesianthropus transvaalensis 31

Porshnev Boris 210, 211, 214

positivism 187

Prezinjanthropus 35

primitive man 190

Proconsul 131

Proconsuls 132

T

Tanzania 34, 35, 38, 55

Tardenoisian culture 162

Taung 29, 30

Tethys 13, 133, 135, 136, 140, 141, 142, 143, 153, 154

thinking 175, 176

Tim White 46

transference 219

Turkana Boy 39, 40

U

Ukhtomsky Alexey 180

unconscious 193, 195, 197

V

visionary activity 231

visions 231, 232

Vygotsky Lev 188, 190

W

Watson John 187

Wundt Wilhelm 177, 194

Z

Zhoukoudian cave 163, 164

9 781916 253216